Caring for Kids

DEDICATION

To Louis A. Petrone (1927–1991) we dedicate this book. Friend, colleague and mentor, Lou helped us formulate this research and gave us good humored encouragement long before any of us could adequately convey our appreciation of him or fully understand what he meant to us. We hope some of his warmth, insight, and sensitivity shows through our words.

Caring For Kids:
A Critical Study of Urban School Leavers

Richard J. Altenbaugh, David E. Engel
and Don T. Martin

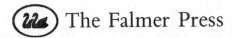 The Falmer Press

(A member of the Taylor & Francis Group)
London • Washington, D.C.

UK The Falmer Press, 4 John Street, London WC1N 2ET
USA The Falmer Press, Taylor & Francis Inc., 1900 Frost Road, Suite 101,
Bristol, PA 19007

First published in 1995

**A catalogue record for this book is available from the British
Library**

**Library of Congress Cataloging-in-Publication Data are
available on request**

ISBN 0 7507 0192 7 cased
ISBN 0 7507 0193 5 paper

Jacket design by Caroline Archer
Cover photograph supplied courtesy of the Pittsburgh Public Schools.

Typeset in 11/13pt Bembo by
Graphicraft Typesetters Ltd., Hong Kong.

*Printed in Great Britain by Burgess Science Press, Basingstoke on paper
which has a specified pH value on final paper manufacture of not less than
7.5 and is therefore 'acid free'.*

Contents

Acknowledgments

This project has incurred many debts and involved many people. Richard Wallace, former Pittsburgh School Superintendent, and Louise Brennan, current Superintendent, supplied an open and candid atmosphere to conduct research in the district. Paul LeMahieu, formerly with the district, helped us design our interview instrument and gain permission from the school district to proceed with this study, and John Barry provided valuable information and services. The good folks at the Pittsburgh Job Corps Center proved to be courteous and hospitable. Vince Moran, the former Director, and Andrea Drozic, the current Director, ensured our access to students and staff cooperation. Job Corps staff included David Bush, Richard Morris and Mardi Amsler, who always were amiable, competent and flexible. We want to thank innumerable counselors for the use of their offices for interview sessions over the years. We too appreciated the opportunity to present our early findings at a regular meeting of the Pittsburgh Job Corps Center's Community Service Group in November 1992. The school leavers we interviewed at Pittsburgh's Job Corps Center taught us a great deal about schooling. Many of them volunteered because, as they articulated it, they wanted to improve the lives of others.

We must acknowledge the program committees for the American Educational Research Association (1993), American Educational Studies Association (1992 and 1993), Association of Teacher Educators (1993), National Dropout Prevention Conference (1992 and 1994), and the Pennsylvania Educational Research Association (1990) for providing us with forums where we could present the products of our research, and receive criticisms and suggestions: this intellectual exchange greatly aided the refinement process.

Without subsidies and other support, we could not have completed this study. We owe a debt to the University of Pittsburgh's School of Education for a Buhl Foundation grant, which subsidized a research assistant, Robert Tazza. Robert's contributions of many interviews, during a nine-month period, helped us to maintain the momentum of this long-range project. Reni Pitts, in the Administrative and

Policy Studies Department, School of Education, at Pitt, spent thousands of hours over an eight-year period transcribing the tape recordings of our interviews; we always treasured her patience and good humor. Debbie Rousseau, a Pitt student, provided valuable research assistance during the early stages of this project. Dean Catherine Morsink, College of Education, Slippery Rock University, provided important grants to subsidize travel expenses for two of Richard J. Altenbaugh's presentations. Her moral support through the latter stages of this project provided additional motivation. Patti Pink, in Slippery Rock's Word Processing Center, spent hundreds of hours typing this manuscript. Her geniality endured through hours of rewrites, revisions, additions and deletions, and impending deadlines. Finally, Jane Flanders, University of Pittsburgh, and Bruce Nelson, Northwestern University, kindly read the entire manuscript, offering many stylistic suggestions. They are in no way responsible for any shortcomings in this study.

We are extremely grateful to Malcolm Clarkson and Ivor Goodson, as well as the staff, at Falmer Press for their persistent support of this study. Without their faith in us, it would not have become a reality. Finally, our families had to patiently endure our physical and mental absences as we conducted our interviews, traveled to conferences, endlessly discussed our findings, and ultimately wrote this manuscript; their support was most precious and invaluable.

Richard J. Altenbaugh
David E. Engel
Don T. Martin

Preface

We began this research because we were impressed, like so many professionals in the 1980s, that school leavers represented a major social and educational problem. The high school completion rate had not substantially increased since the mid-1960s. In some areas, mostly urban centers, the school leaver rate was growing worse. Students had seemingly not improved in national reports of Scholastic Aptitude Test (SAT) performance. Business leaders in the United States bemoaned the linguistic and computational skills of the typical public high school graduate entering the workforce. It all seemed to be pretty grim. How was the nation going to compete internationally if our high school graduates could not perform up to standard with their Japanese and European counterparts? If 25 to 30 per cent of high school students were early school leavers, did this not create a growing number of unemployable youth?

I was especially concerned because, in the early eighties, I was a member, and for two years President, of the Board of Education of the Pittsburgh public schools. During those years the dropout rate fell from the low 30s to the mid 20th percentile. I consulted the Superintendent and the Director of Research and Evaluation seeking an explanation. To what could this apparent improvement be attributed? The Director of Research could cite statistics, but comparisons to other similar school districts were difficult to make because different states computed dropout rates in different ways. The Superintendent could offer some intuitive hunches, but neither had data from the students' experiences. Both agreed that documentation of students' reasons for leaving school before completion would be useful. So the concept driving this research was born.

It was obvious from the outset that actual dropouts would need to be interviewed. An interview protocol was developed in consultation with the district's Director of Research and Evaluation and related literature was reviewed. Several diversions got in the way, not the least of which was contacting subjects for interview.

Initially we obtained a list of forty-eight students who had dropped out of Pittsburgh high schools. In all but one case, we found that the dropouts were no longer at the address on the list. Where had they gone? No one could tell us: it was if they had fallen off the face of the earth. Still, it was necessary to test the interview protocol before proceeding to a larger sample. The one student who seemed to be at the last known address on the list was called. He agreed to an interview. On the appointed day and hour, that student failed to appear; with a follow-up telephone call, I was told by that motherly electronic voice that the number had been disconnected. We seemed to have reached a dead end.

Then Don Martin came up with a fruitful suggestion. He had prior experience with the Pittsburgh Job Corps program and knew that it served school dropouts who wanted to obtain a Graduate Equivalency Diploma (GED). Here was an accessible group to interview. After making arrangements and gaining clearance from the Job Corps and the US Department of Labor, which oversees the program, we began interviews.

The students we interviewed were at variance from those we initially had anticipated. Because they had taken the initiative to obtain a GED, they could not be viewed as leaving school in any absolute sense. They were motivated to return to an alternative educational program. In most cases, they went to the Job Corps to learn a vocational skill as well as earn the GED. They have been able to tell us why they left school as well as why they dropped back, thus enriching our data; we refer to them as *dropbacks*. They stopped their schooling and then restarted, showing us that there can be life after leaving school.

This experience has renewed our faith in American education. Instead of believing that the rate of school leavers catalogues failures, we now feel that there is potential for students who leave school early. If those who leave school before completion can find alternatives to ordinary programs, they can rejoin their peers with a high school diploma.

David E. Engel
Pittsburgh, PA, 1994

Part I

Introduction

Chapter 1

The Social Terrain

No one really knows what causes students to drop out of high school. (Rumberger, 1986)

Studies of school leavers, relying mainly on statistical approaches and descriptive narrative, too often produce ill-informed policies. This literature generally remains ahistorical, viewing school leaving as a wholly contemporary issue. It also lacks a social framework, avoiding critical analyses. In contrast, this study attempts to place our student narratives within broader historical, philosophical and social contexts. Our purpose is only partly utilitarian. It is committed to understanding the experience of school leaving: we reconstruct schooling through students' perceptions in order to gain some insight into the school leaving process.

From a traditional perspective, the United States appears to face a deep and nagging dilemma, which seems even more problematic given current trends. 'While national estimates of rates of leaving school before a diploma range from 18 to 25 per cent of 18-year-olds, estimates from large cities are often double these rates, and, for some subgroups of urban students, rates have been reported at 60 per cent or higher' (Hammack, 1986, p. 326; see also Mann, 1986, p. 311). In certain localities, Hispanics claim a 78 per cent rate, with Native Americans as high as 90 per cent (Kunisawa, 1988, p. 62). Rumberger (1983), pointed to yet another disturbing pattern: the school leaver rate appeared to be escalating among white middle-class youth (p. 200). Asian-American students, with a 9.6 per cent school leaver mark, represent the only exceptions (Kunisawa, 1988, p. 62). Finally, slightly more males than females leave school, at 53 and 47 per cent, respectively (Beck and Muia, 1980, p. 66; Markey, 1988, p. 37). It appears from all accounts, therefore, that the population at-risk transcends race, ethnicity, social class, as well as gender (Natriello, Pallas, and McDill, 1986; Stoughton and Grody, 1978). Moreover, although this country has experienced a long-term decline in the incidence of school leaving, 'the short-term has remained steady and even increased' (Rumberger, 1986, p. 101; General Accounting Office, 1987, p. 10).

Full attendance has never been achieved, and ironically recent school reform policies may exacerbate, rather than ameliorate, this problem. An avalanche of reports, issued amid heavy media blitzes since 1983, condemned the general 'crisis' in American public education and have, with almost total unanimity, called for a more academically oriented curriculum and tougher graduation standards. High schools have received special attention in this regard (Boyer, 1983; Goodlad, 1984; National Commission of Excellence in Education, 1983; Sizer, 1984). With much fanfare and for obvious political purposes, George Bush in 1988 declared himself the 'education president'. He anointed, as well as reinvigorated, this movement when he proclaimed in his 1990 State of the Union message: 'The nation will not accept anything less than excellence in education' (quoted in Spring, 1991, p. 23). Bush and the National Governor's Association established six basic goals to be attained by the year 2000, calling for 'competency in challenging subject matter' and striving for 'first in the world in mathematics and science achievement' while at the same time prescribing a 90 per cent high school graduation rate. Showing little imagination, the Clinton administration, with its 'Goals 2000: Educate America', has chosen to maintain the same basic objectives. National assessment, based on standardized test scores given in the fourth, eighth and twelfth grades, represents the means to measure and report progress ('Text of Statement', 1990, p. 16; 'Tracking Progress', 1991, p. 6; Spring, 1991, p. 23; 'Riley Announces "Goals"', 1993, p. 1).

However, simply raising academic codes and instituting more tests could leave more students 'behind in the pursuit of excellence' (Natriello, 1986, p. 306; Alexander *et al*, 1985; Foley, 1985; Natriello, Pallas and McDill, 1986). Assessment, which often becomes 'the primary form of education reform', masks this deep-seated problem: 'Testing . . . can only measure progress . . . not engender it . . . Assessment would not address the issue of rigid and bureaucratic school governance and structure, high school dropout rates, teacher quality, or a whole host of other issues critical to school reform' ('National Testing Debate', 1991, p. 2). This school reform movement appears to have some degree of failure, rather than success, built into it by further aggravating the 'fragility of school completion' (Mann, 1986, p. 310), because increased standards through measurement alone could result in frequent grade repetition, which would further undermine attendance. 'Being retained one grade increases the risk of dropping out by 40–50 per cent, two grades by 90 per cent' (Natriello, 1986, p. 308; Voss, Mendling and Elliot, 1966, p. 365).

Research on school leavers unfortunately has not provided policy

makers with adequate answers to this problem, and has at best produced ambiguities. In spite of a large number of studies, they appear
to be plagued by three widely varying but significant obstacles: definitions, causes, and solutions.[1]

Definitions

Until recently, definitions differed, and at times confused the issue
(Hammack, 1986, p. 328; Morrow, 1986, p. 343; Rumberger, 1986,
p. 103; General Accounting Office, 1987, p. 38; Stoughton and Grady,
1978, p. 312). In 1987, a frustrated Government Accounting Office
(GAO) report observed that only forty-one states even bothered to
'count students who drop out of school and that states vary in their
definitions of a dropout. One state includes a transfer to a non-public
school, thirty-four states include military enlistees, twenty-one states
include persons completing a GED, eight include education-at-home
students, and thirty-two include expelled students' (General Accounting Office, p. 40). Consequently, *pushouts*, or expelled students, received
inconsistent reporting (Mann, 1986, p. 309). School leaving criteria
fluctuate as well: 'Eleven states use the lack of a school transcript as
a factor that classifies a student as a dropout' (Hammack, 1986, pp.
327–8).
  Structural questions have also clouded this issue. Special schools
and alternative programs as well as different grade sequences have
perplexed researchers and federal agencies alike. An exasperated
Hammack (pp. 327–8), in a 1986 survey of urban school leavers, that
focused on Boston, Chicago, Los Angeles, Miami, New York and San
Diego, noted that 'some districts include special education students in
their reports, while others do not; some include all students enrolled in
any type of program offered by the district, while others include only
those enrolled in regular day high schools'. Grade levels also lacked
uniformity, according to the 1987 GAO study (General Accounting
Office, pp. 40–1): 'Among forty-one states, twelve states report dropouts for grades 9–12 and fifteen states count grades 7–12, with most of
the others (twelve) reporting dropouts for kindergarten through grade
12'. Of course, the accuracy of these statistics depended on the thor-
oughness of 'centralized record keeping' (Hammack, 1986, p. 327).
This included, as the GAO (1987), lamented, 'time periods during the
school year that dropout data are collected' and 'tracking or follow-up
of youth no longer in school to determine if they continue or complete
secondary education elsewhere' (p. 40). Because of these unreliable

reporting procedures, students simply disappeared in this Kafkaesque, bureaucratic maze.

Recent attempts by the Council of Chief State School Officers (CCSSO), with support from the National Center for Education Statistics (NCES), to introduce uniformity in the definition of school leavers and in reporting practices, so as to gather 'accurate statistics', have also proven to be frustrating (Clements, 1990, p. 18). Selected school districts in thirty-one states have been piloting a reporting system developed by the CCSSO and NCES. The base population now includes special students, or those in alternative public-school programs, and compulsory school-age youths who have not graduated. This definition excludes school-age children in 'prisons, mental institutions, juvenile institutions, and adult training centers' from the base population (*ibid*, p. 21). Reporting procedures involve annual school leaver reports and cover grades 7 through 12. Early returns from this pilot project appear mixed, however. States still vary regarding expulsions and home-based instruction, and transfers remain difficult to track, particularly those to non-public schools. Some school administrators have also complained about the 'time and expense involved in revising current collection and reporting practices to meet the proposed national standards' (Goldman, 1990, p. 20). A few Wisconsin districts have likewise complained about their participation because no uniform transcripts or student identifications exist below tenth grade. On the other hand, Florida and Mississippi districts have found the new definition and reporting procedures usable and enlightening. In the end, no consensus exists even among the pilot programs, which further confuses the issue of defining school leavers.

Worse yet, and as alluded to above, most districts do not report, and too many researchers fail to study, students who return to school, otherwise known as 'dropbacks'. Approximately 10 to 33 per cent of all school leavers return, and 90 per cent of these continue their education onto post-secondary levels. 'Some do not rejoin high school but try another sort of post-secondary institution' (Borus and Carpenter, 1983, p. 501; Mann, 1986, p. 315; Weis, Farrar and Petrie, 1989, p. x). These returnees tend to be young and single, who long maintained post-secondary aspirations. As Borus and Carpenter (1983) conclude, one set of variables shaped their decisions to terminate schooling while another set effected their resumption: 'For the most part, return to school seemed to be based on the individual's characteristics rather than on school-related factors' (pp. 502–3). We believe that their perspectives of why they left school and then returned would prove invaluable to educators and policy makers.

Causes

Investigating the reasons for school leaving poses an equally perplexing problem. Many excellent school leaver studies, especially Project TALENT (see Bachman, O'Malley and Johnston, 1979), Youth in Transition, the National Longitudinal Survey of Youth Labor Market Experience (see Borus and Carpenter, 1983; Rumberger, 1983), and High School and Beyond (Kolstad and Owings, 1986, p. 7), have produced extensive data that review the reasons for school leaving, and made 'several generalizations' about them (Wehlage and Rutter, 1986, p. 375). School leavers, according to these reports, come from impoverished families, accrued feeble academic records, maintained high failure rates, and reflected poor outlooks, such as 'negative school attitudes, low self-esteem, and external locus of control' (*ibid*; Beck and Muia, 1980, p. 66; Boyer, 1983, pp. 244–5; Kowalski and Cangemi, 1974, p. 41; Rumberger, 1986, p. 109; Stoughton and Grady, 1978, p. 314; Wagner, 1984; Weidman and Friedmann, 1984, p. 27).

More important, these explanations have remained somewhat static for decades (Rumberger, 1983, p. 201). Sherman Dorn (1993, pp. 356 and 363) argues that the modern school leaver dilemma assumed legitimacy and gained wide public attention beginning in the 1960s, with the publication of several hundred articles in education journals. The manpower concerns of the 1950s, and demographic changes, i.e., 'more teenagers graduated from high school', set the stage for the sixties' explosion in research into the school leaver phenomenon. School leaving emerged as a deviant activity, as Dorn (1993) explains:

> Much space within the dropout literature was devoted to five motifs, all of which were to some extent explicit: equating the dropout problem with unemployment, linking it with urban poverty, using the language of juvenile delinquency, assuming that dropouts were male, and asserting that psychological defects were a primary distinction between dropouts and graduates. (p. 363)

These themes persisted, evident in contemporary 'public debate'.

Such categories, of limited help, tend to oversimplify matters and, obfuscate intricate patterns. Melissa Roderick (1993, pp. 43 and 82) attempts to unravel the process by concisely analyzing the school leaver literature and placing it into three groups. First, at one end of the spectrum, many studies point to student background as primarily responsible for leaving school, diminishing the impact of schooling.

Second, at the other end of the spectrum, some literature focuses solely on school 'structure, organization, and policies'. Third, more complex research stresses a combination of factors, i.e., family background as well as insensitive and inadequate school policies and staff. 'The singular outcome — not finishing high school — is in fact a nest of problems' (Mann, 1986, p. 311; Wehlage *et al*, 1989, pp. 25–6). We fall into this latter category, yet place more emphasis on the school's role in the leaving process; we cannot — nor do we presume to — change family conditions, but we can recommend alterations in school culture and structure to mitigate school leaving.

The methodology used to conduct school leaver research often contributes to its shortcomings. Descriptive statistical studies provide valuable information, to be sure, but generally lack in-depth analysis of this social and educational problem. While Rumberger (1983, pp. 210–11), for example, sees family background as strongly influencing 'the probability of dropping out for members of all race and sex groups', he cautions that 'as with all previous studies of dropout behavior, the results obtained from these models have certain limitations'. These research efforts, he continues, present 'associations between independent variables and the probability of dropping out', yet they do not 'infer causality from various factors'. These 'factors' might really be 'symptoms' rather than causes of dropout behavior (see also Bachman, O'Malley and Johnston, 1979, p. 482). Hence, these studies fail largely to reveal the mechanisms that actually caused students to abandon schooling. As Rumberger (1986, p. 109) concludes in another study: 'No one really knows what causes students to drop out of high school'. Even worse, we know little, virtually nothing, about the *process* of student disengagement.

Most studies also fail to account for the causes of why school leavers resume their schooling. This represents a serious oversight, because, as Borus and Carpenter (1983) ironically contend, returning to school defies most traditional reasons for leaving it:

> Family background variables, including father's education, poverty status, and absence of mother and/or father in the home at age 14, all which increased the probability of dropping out of school, seemed not to alter the rate for returning. Likewise, the proxy for ability, knowledge of the world of work score, was not significant. . . . Having had or parenting a child . . . was not a significant factor after marital status was accounted for. This implied that it is marriage rather than a presence of a child that hinders returning to school. Finally, the local unemployment

rate, personal unemployment status, and being from a poverty household were also not statistically significant, which would appear to indicate that economic conditions do not induce drop-outs to either remain in the labor market or return to school when the other factors are controlled. (p. 505)

Their statistical analysis of this phenomenon is helpful but limited, because, like most investigations, these researchers examine the results of school leavers' decisions, not the decision-making process. And this leads us to our central point.

School leaver research typically dismisses the students' perspective. Of course, the ambitious National Longitudinal Survey of Youth Labor Market Experiences relied, in part, on data obtained through 'a series of annual interviews for a national sample of approximately 12,700 young men and women', while the large-scale High School and Beyond study utilized questionnaires (Rumberger, 1983; see also Borus and Carpenter, 1983). Yet, as Wehlage and Rutter (1986), point out in their valuable analysis, 'although the major studies sought student views, there is a tendency by researchers to see such information as less important, or at least to treat it as "surface" data as opposed to "underlying" data, which are assumed to be more "powerful"' (p. 376). Except for surveys and questionnaires, students, who are the principal actors in the school leaving process, have been treated, at best, marginally and, at worst, overlooked.

Some recent studies depend on interviews with school leavers. Such an approach defies seemingly neat patterns, shedding light on the process as well as the causes of leaving school. As Farrell (1990) convincingly argues: 'There is a myriad of statistical information available on the dropout phenomenon with which educators have attempted to go from the general to the particular. To get the at-risk students' view, however, we have to do the opposite — go from the particular to the general' (p. 6). His examination, which grew out of the New York City's Stay-in-School Partnership program, relies on ninety-one student interviews. Its analysis, while insightful, seems limited because it concentrates on a psychological perspective to explain dropout behavior, tapping Erik Erickson's concept of adolescent 'self' with its 'conflicting selves' (p. 3). A broader context is needed to better understand the social forces and educational conditions shaping dropout behavior.

Fine's (1985, 1986 and 1991) ethnographic research focusing on a New York City public high school, uses interviews of administrators, teachers, and students, the latter supplemented by surveys, to demonstrate social 'reproduction': 'The analysis relies upon life in this school as a way of examining how the act of dropping out, even if intended as an act of social resistance, ultimately reproduces and exacerbates social inequities' (Fine, 1985, p. 44). However, she only focuses on one high school, limiting generalizations.

Overlooking students' perspectives can also diminish the importance of schooling itself as a cause of leaving. According to Wehlage and Rutter (1986, p. 376), 'there is a clear trend in what students say. They leave because they do not have much success in school and they do not like it. Many of them chose to accept entry-level work to care for their children, choices that apparently are seen as more attractive than staying in school' (p. 376).[2] This is what Ken Reid (1983) found in his study of school absenteeism in South Wales. He interviewed 128 persistent school absentees, selected from two inner-city comprehensive schools in an industrially depressed area, in order to gain insights into their initial and continued reasons for missing school. His findings suggest that despite the absentees' generally unfavorable social and educational backgrounds, a greater proportion of these students seemed inclined to blame their institutions rather than psychological or social factors for their behavior.

Solutions

This kind of information is critical because research approaches and findings shape the solution. 'The focus on social, family, and personal characteristics does not carry any obvious implications for shaping school policy and practice. Moreover, if the research on school leaving continues to focus on the relatively fixed attributes of students, the effect of such research may well be to give schools an excuse for their lack of success with the dropout' (Wehlage and Rutter, 1986, p. 376). Such cynicism fuels the 'blame-the-victim' perspective (Mann, 1986, pp. 310–11). Wehlage and Rutter (1986) warn against such shortsighted, traditional lines of research, arguing for new approaches: 'Researchers need now to ask why these youth are educationally at risk and, further, what policies and practices of public schools can be constructive in reducing the chances that these students will drop out' (p. 377).

Countless commissions have been convened 'to address the problems of at-risk students' (Rumberger, 1986, p. 116). Past as well as present policy recommendations have usually followed standard causal research data. Current efforts 'that seem to work' include 'work-experience programs', small-scale settings that emphasize caring, computer-assisted instructional techniques, computerized monitoring of students at risk, and 'business-school partnerships' guaranteeing employment for high school graduates (Mann, 1986, pp. 318–20; Bachman, O'Malley and Johnston, 1979, p. 482; Balfour and Harris, 1979; Wagner, 1984; Weidman and Friedmann, 1984, p. 37). Other policy makers and researchers have pointed to a variety of alternative schools or programs (Boyer, 1983, pp. 245–6; Farrell, 1990; Foley, 1985; Kunisawa, 1988). Finally, the characteristics of 'reentry' programs have differed markedly from prevention efforts (School Dropouts, 1987, pp. 22–5).

Nevertheless, without the students' perspectives, we simply do not know what will work (*ibid*, p. 20). As Wehlage and Rutter (1986) maintain: 'From the standpoint of school policy and practice, it is essential for educators to become knowledgeable about the way school can be perceived differently and can affect different groups of adolescents in different ways' (p. 380).

A New Direction

Other studies and findings directly confront the flurry of publicity over the 'educational crisis' in general and the school leaver dilemma in particular. Bracey's findings are provocative, since he sees no decline in American schools; rather, just the opposite: The 'education system — as a system — continues to perform better than ever' (Bracey, 1992, p. 107). Concerning the specific school leaver problem, Bracey (1991) states that 'high school graduation rates are at an all-time high' (p. 106). Not only did the United States experience an 83 per cent graduation rate in 1989, but this represented a 'misleadingly low' figure because it only accounted for those students who graduated within the traditional twelve-to-fourteen-year period (*ibid*). Unlike the situation in many other countries, growing numbers of young Americans resume their schooling, often completing it. 'In 1989, 87 per cent of Americans between the ages of 25 and 29 held high school diplomas or GED (general equivalency diploma) certificates, up from 73 per cent only twenty years earlier' (*ibid*, p. 107). Even more startling, and undermining the credibility of the 'Goals 2000: Educate America' graduation projection, '91 per cent of the class of 1980 had completed high school or its equivalent by 1986' (*ibid*; NCES, 1994). The typical twelve-year template does not apply to all students, due to a number of variables including development, maturity, and life experiences. The standard expectation is understandable, yet it masks the fact that some do complete school over a longer period of time and often in unconventional ways. Thus, when young people take fourteen to sixteen years to finish school, the leaver situation appears to be significantly different.

Return rates therefore prove encouraging. However, dropbacks — those dropouts who resume their schooling — receive little or no attention from researchers. According to Kolstad and Owings (1986) who tapped the High School and Beyond data, 38 per cent of school leavers nationwide 'returned and completed high school or obtained a GED' (p. 14). This resumption rate varies depending on when students first abandoned schooling, as well as their gender, race, ethnicity, social class, test scores, location, post-secondary expectations, and employment opportunities. Twenty-seven per cent of sophomore school leavers returned to complete their graduation requirements, with 37 and 41 per cent for juniors and seniors, respectively (*ibid*, p. 15). Although slightly more males than females leave school before graduation, female school leavers returned at the same general rate as males. However, when Kolstad and Owings combined gender with ethnicity and race, different patterns emerged: 'Among majority whites, young male and female dropouts were about equally likely to return and complete high school, but among Hispanics and Blacks, young male dropouts were about 10 percentage points more likely to return and complete high school than young female dropouts' (*ibid*, p. 16).

These researchers also isolated ethnic and racial differences: 30 per cent of Latino, 33 per cent of African-American and 41 per cent of White school leavers returned. Yet socioeconomic status significantly affected these figures. Affluent students, regardless of their ethnic and racial backgrounds, returned at a higher rate, with 42 per cent for Latinos, 44 per cent for African Americans and 56 per cent for Whites. These rates dropped dramatically for low-status students, with 32, 25 and 32 per cent respectively (*ibid*, p. 27; Weidman and Friedmann, 1984, p. 26). When Kolstad and Owings (pp. 17 and 28) introduced academic achievement, as measured by standardized test scores, they found yet another pattern. In the upper three test score quartiles, 69 per cent of Latino and 58 per cent of African-American school leavers returned, with 55 per cent for Whites. These percentages plummeted to 18, 25 and 22 per cent, respectively, for low-achieving students.

Location too affected the dropping back process. 'The South and Northeast had return/completion rates around 40 per cent, compared to a 35 per cent rate in the West and North Central regions'. Dropback rates too varied according to the community: 'High school dropouts in urban areas had dropout/return rates around 35 per cent, compared to 37 per cent in rural areas and to 42 per cent in suburban areas' (*ibid*, p. 18).

Educational goals likewise influenced return rates, as Kolstad and Owings point out:

> Those who expected to go to college, but dropped out of high school, are more likely to return and complete high school than those dropouts who had no further educational plans for after high school (61 per cent compared to 27 per cent). Among those who had an intermediate level of educational expectations (junior college or vocational/technical school, male dropouts were more likely than female dropouts to return and complete school (51 per cent compared to 44 per cent for those who expected vocational/technical training), and 64 per cent compared to 46 per cent for those who expected to attend junior college). (*ibid*, p. 19)

Finally, employment produced fascinating, but ambiguous, results. Seventy-nine per cent of employed male school leavers remained 'stayouts', not completing high school or a GED. The experience of the employed female school leaver contrasted with this; 51 per cent eventually returned to school. Regardless of the variables, school leavers nationwide believed that leaving school was a mistake. Kolstad and Owings (*ibid*, pp. 3, 18 and 20) conclude their study on a positive note, seeing viable solutions, such as reentry programs, to the school leaver dilemma.

Bracey (1991) too recasts the minority experience: school leaver rates appear to be 'declining for all ethnic groups except Hispanics, for whom the rate is steady' (p. 107). Poor and inconsistent data-gathering techniques, as mentioned earlier, have somewhat distorted perceptions: 'In some reports that calculate dropout rates, a *dropout* is any person without a high school diploma who is not in school. Thus many undereducated *adults* who have immigrated recently from South and Central America are labeled as dropouts from a system that they never entered' (*ibid*). Our biases toward school leavers likewise need to be revised, as Bracey stresses: 'Contrary to the popular stereotype of dropouts as largely Blacks and Hispanics, 66 per cent of dropouts are White. Sixty-eight per cent come from two-parent families, 42 per cent come from suburban high schools, 71 per cent never repeated a grade, and 86 per cent live in homes where English is the native language' (*ibid*). This mitigates the notion, which we pointed out above, that poverty and non-traditional family structure 'cause' the school leaver phenomenon. Nevertheless, Bracey qualifies this by recognizing the difference between 'rates' and 'numbers' (*ibid*). African-American school leaver *rates* certainly overshadow those for Whites, while Hispanic *rates* outstrip most groups, 'but the *number* of White dropouts is much larger because Whites still represent the dominant population group in the country'.

Yet even aggregate school leaver rates, if not handled carefully, can prove misleading. Recent research has shed a different light on this issue. Kominski (1990) approaches the school leaver rate from a purely statistical standpoint, and finds the usual serious inconsistencies in data gathering techniques which have distorted this phenomenon. He analyzes annual school leaver rates, focusing on secondary grade levels, and tracks it between 1968 and 1985. He concludes:

> At the national level, there is little difference between the rates of 1985 and 1968. The data show that substantial improvements have been made by Blacks since 1968, but White rates, already low in 1968, have experienced no overall improvement. Rates for Hispanics, available only since 1972, show much higher levels of dropping out at all four high school grades. (pp. 308–9)

Rumberger (1991, pp. 67–8) reinforces this last point, focusing on Chicano students. 'At virtually every age group, dropout rates are higher for Chicanos — roughly twice that of Whites and higher than any other ethnic or racial group except American Indians'. Nevertheless, while Bracey (1991) avoids trivializing or minimizing the 'problem of high school dropouts in the United States' (p. 308), he brings a sober perspective to it by addressing distortions and overstatements, as well as exposing its complexities.

These perspectives reflect increasingly popular attempts to offer hope amid the gloom of purposeful 'disinformation' during the 1980s (Bracey, 1991; Bracey, 1992, p. 108; see O'Neil, 1993).[3] These views also acknowledge the oversimplification of educational problems, like school leaving. It should be noted that any rate of school leaving remains unacceptable. However, the recent political and media sensationalism over this nation's school leaver situation has grossly exaggerated and obfuscated an extremely complex and variegated phenomenon. Politicians, teachers, school administrators and the general public do not have a clear picture of school leavers' experiences. We argue that the school leaver dilemma is not as bleak as many would have us believe; our findings in fact reveal a somewhat hopeful situation. This does not diminish the fact that our society should strive to ensure that all students complete their initial schooling. In lieu of this, we maintain that with effective local, state and federal prevention and reentry programs the adverse social effects caused by students leaving their schools will continue to decline.

The Pittsburgh Study

This study concentrates on school leavers in Pittsburgh, a large metropolitan center. Once the world's leading steel producer throughout most of the twentieth century, the 'Steel City' now symbolizes what has now become the American 'rust belt'. Rather than a steel center, it has spent the past decade struggling to become a leading corporate and service center. During this difficult transition, from the late 1970s to the mid-1980s, the school leaver trend in the city's schools contradicted the literature that indicated a nationwide increase in school leavers. According to the administrative records of the Pittsburgh public schools, which maintained data on students who leave after the age of compulsory attendance and prior to graduation, the school leaver rate had fallen from the low 30th percentile to the low 20s. However, a report released in 1986 indicated a sudden surge in the rate, from 21 to 27 per cent, during the previous academic year (*Pittsburgh Post Gazette*, 17 December 1986). Descriptive statistical approaches alone could not explain why students decided to abandon the Pittsburgh schools at this, or any other, time. This policy-oriented study, focusing on school leaver prevention and resumption, explores four basic questions: Who left school? Why did these students leave school? What caused them to return? What intervention policies can be formulated to prevent students from leaving school? We believe the results of this research on one city will have wider implications, as we place them within broader historical, social, and philosophical contexts.

We interviewed 100 *dropbacks* enrolled in Pittsburgh's Job Corps Program during an eight-year period in order to explore school leaving and resumption experiences. Since these students came from all of Pittsburgh's high schools, we have a composite picture of the entire system. Our research approach was designed to maximize students' responses from their own frames of references, not from a closed set of prearranged questionnaires or surveys. They reflectively reconstructed their school experiences for us as we sat with them for countless hours at the Pittsburgh Job Corps Center. Their recollections and insights informed us and sometimes surprised us.

We rely on these in-depth, oral interviews of *dropbacks* to understand why they left school and why they returned, to reconstruct their subjective meanings of schooling. This approach stems from symbolic interaction, as Bogdan and Biklen (1982) summarize it:

> it is not the rules, regulations, norms or whatever that are crucial in understanding behavior, but how these are defined and used in specific situations. A high school may have a grading system, an organizational chart, a class schedule, a curriculum, and an official motto . . . People act, however, not according to what the school is supposed to be, or what administrators say it is, but rather, according to how they see it. (p. 34)

As with such field studies, our theoretical framework is grounded in the data, thus inductively analyzing the findings. 'Theoretical constructs might be useful, however, they are relevant to understanding behavior only to the degree that they enter in and affect the defining process' (*ibid*).

Our sole concern has been, from the beginning of this project, to capture school leavers' perspectives, to focus on the 'dropout rather than the valedictorian'. 'Success' marks only one side of schooling; 'failure' exists as well (Grumet, 1988, pp. 59 and 67). As Rumberger (1991) warns about the complexity of this issue:

> We use dropping out as a visible and convenient measure of academic failure and graduation as a visible and convenient measure of academic success when neither reveal much about how much or how little knowledge a student has acquired. Thus, in some respects, too much attention is being placed on dropping out and graduating, when we should be more concerned with student engagement, learning, and knowledge. (p. 67)

Some readers may see our emphasis as distorting reality by only interviewing school leavers rather than administrators, teachers, counselors and even parents. Why? Is this study any less credible because it only includes the views of our schools' and society's 'dropouts'? Our approach features voices seldom, if ever, heard. School leavers' perceptions represent their realities, and these perspectives are as legitimate as any other group's perceptions, if not more so. Their realities shaped their decisions to abandon schooling and then to resume it. We must understand and value this process and these voices. If we do not, we will never resolve the school leaving dilemma.

The structure of this study highlights their perceptions, and analyzes them against a broad backdrop. Part I supplies a contextual framework. This opening chapter provides a synopsis of the literature, synthesizing it around definitions, causes, and solutions, to demonstrate how this investigation addresses major lacunae. Chapter 2 examines school leaving against a historical backdrop in order to gain a sense of continuity through human agency. Children and their families have made choices about education since the advent of widely available public schooling in the nineteenth century. That chapter emphasizes the conflict, rather than the consensus, of authority over children's welfare, the dispute between compulsory attendance and family rights; it also explores the impact of the emergence of the 'adolescent' as a new phenomenon on high school completion. Chapter 3 begins by briefly looking at Pittsburgh's historical, economic and social contexts, with a short history of the Pittsburgh school system and historical trends. It draws on the classic and rich Pittsburgh Survey, conducted and published in 1911 by the Russell Sage Foundation. That chapter continues by elaborating the methodological approach and describing the narrator population; often articulate and always thoughtful, these former students usually defied the stereotype of the school leaver.

Part II concentrates on student perspectives of schooling and why they first decided to abandon it and why they then chose to resume it. The school setting serves as the topic for chapter 4. Narrators describe the physical plant, often overwhelmingly large and chaotic, and the social scene, usually marginal. Chapter 5 portrays school personnel through students' eyes, from building administrators, like principals, vice principals and deans, to teachers, security guards and custodians. On the one hand, they categorized administrators as invisible at best and adversarial at worst. On the other hand, our interviewees consistently pointed to teachers as the key 'persons' in their decisions about learning as well as their attitudes toward schooling. School knowledge dominates chapter 6. We were surprised to hear former students identify the curriculum with teachers, that is, whether a student liked or disliked a particular subject depended on a teacher's personality or pedagogical skill; knowledge did not stand alone, but represented a humanistic process. Chapter 7 views the school leaving process as complex, and grapples with the unique experience of 'dropping back', which was seen as seldom calculated, sometimes serendipitous, and always necessary.

Part III analyzes these findings and offers policy recommendations. Chapter 8 attempts to explain why students saw their school leaving and dropping back based on solely individual considerations and perspectives. For instance, few of our narrators, though largely minority students, reported systemic racism; yet they did point to isolated racist incidents with teachers. Still, they rarely blamed racism for their decisions to abandon schooling; rather, many of these students saw their leaving as an isolated act, for which they later expressed regret. Labor market theory represented an important explanation for their decisions to return. The humanistic side of students' comments comprise chapter 9. They voiced a sense of marginality and often characterized school personnel as uncaring. Chapter 10 concludes by reviewing a variety of existing intervention and reentry policies for at-risk students and how our investigation supports or refutes them. In our case, for example, we found that a federal reentry program like Job Corps works. This confronts the ongoing reform movement which, during the Bush administration, diminished the role of the federal government, emphasizing solutions that had been created and subsidized at the local levels. Such shortsightedness will only exacerbate the school leaver problem.

Finally, we prefer to use the broad term *school leaver* (Wehlage, 1989, p. 1). This represents an all-encompassing label, which includes dropping out, pushing out, fading out, easing out, as well as combinations of these various school leaving experiences. *School leaver* also avoids negative, almost pathological, connotations associated with the *dropout* label. *Dropout* implies that something is wrong with the individual, when in fact home or school experiences, peer pressures, or social or economic demands may have shaped their decisions; nevertheless, leaving school remained their decisions. We therefore see students as human agents, responding to various conditions and making choices — albeit, at times, limited ones — about first abandoning and then resuming school, or undertaking some educational equivalent.

Notes

1 A special issue of *Teachers College Record*, spring 1986, provided sharp focus for this study. We were grappling with these very same matters at that time. Rumberger (1986) maintains a similar structure, i.e., definitions, causes, and solutions, in his review of the literature.
2 See also Kunisawa (1988): 'The problem is not the dropouts; the problem is a dysfunctional educational system that produces dropouts' (p. 63).
3 Bracey's contentions have not gone unchallenged. Refer, for example, to Stevenson (1993).

Chapter 2

School Leavers in American Society

What ye learn in school ain't no good. (Child Laborer, 1912–13)

School leaving represents more than a late twentieth-century phenomenon, existing since the inception of the common schools in the nineteenth century and circumventing compulsory education laws. History
sheds a very different light on this experience. In 1900, 90 per cent of
students did not complete high school; by 1940 this figure had only
decreased to 76 per cent. School leaving did not fall below 50 per cent
until the 1950s, and reached its lowest point during the sixties, with
12 per cent in 1967. By 1970, however, that figure rose to 17 per cent,
rigidifying, according to some interpretations, at 25 per cent in recent
years (Kolstad and Owings, 1986, p. 11; Kunisawa, 1988, p. 62; Mann,
1986, pp. 311 and 313; Wehlage and Rutter, 1986, p. 374; Wehlage
et al, 1989, p. 30).

American educators have long expressed frustration over the act
of school leaving. In 1872, William Torrey Harris, St. Louis School
Superintendent, delivered a rambling paper at the annual meeting of
the National Education Association (NEA) in which he outlined his
impressions of the causes for early school leaving. First, the existing
school experience lacked systematic early childhood education, namely
at the kindergarten level, which in turn failed to encourage students 'to
love school, and to form good habits'. Second, harsh teacher discipline, in the form of corporal punishment, alienated many students
who escaped humiliation and pain by abandoning school. Third, the
lack of formal and widespread age grading created deep inefficiencies,
failing to challenge the 'best pupils' and overwhelming the 'poorest
pupils' (Harris, 1873, pp. 265 and 266–7). Nevertheless, the actual
term *dropout* did not appear until 1900 (Tyack and Hansot, 1990, pp.
166 and 330) when James Greenwood (1900, pp. 341 and 343), Superintendent of the Kansas City schools, who had collected statistics from
fifteen urban districts in order to calculate the number of 'dropouts'
and explain why they 'quit high school during the first year', presented
his findings at the 1900 NEA meeting. Age appeared to be the most

consistent variable; only 11 per cent of 12-year-olds abandoned school, while 60 per cent of the 18-year-olds abandoned high school during their first year. A. Caswell Ellis (1903), a University of Texas education professor, followed three years later with another NEA school leaver report. In a clearly disturbed tone, he criticized high schools for only graduating 'about 2 per cent of the original number entering the elementary schools' (p. 793). He pompously invoked a military image to dramatize his point: 'The slaughter of the Light Brigade at Bala Klava pales into insignificance, then, beside the slaughter of the educational hopes and possibilities of our children by the present school system'. He focused on boys because they compiled the poorest attendance and completion records. He also expressed frustration about gathering data. He cited Greenwood's study of superintendents, various principal reports, and scattered teacher comments, but questioned the reliability of these 'second-hand child' investigations. Ellis complained to those assembled: 'If one could get two thousand boys who left the high school to tell frankly why they left . . . we should have a contribution to child study of real value'.

Almost 100 years later, our young narrators both fascinated and puzzled us with their strong sense of individualism. They saw their school leaving and dropping back experiences as tied to no other factors than their own decisions. Family, school, and social conditions certainly shaped their choices, but these students consistently saw their actions based on self-direction. This strident evocation of individualism confused us at first, since we expected them to blame their family, school, or society. Such was not the case. How could we explain this? We turned to a broad historical analysis of youth culture as one means to shed some light on this phenomenon, asking two basic questions: What have been the patterns of school leaving over time? More specifically, how have the changing relationships between children, families, and schools unfolded?[1]

Three concepts, compulsory schooling, family economy and adolescence, focus this contextual analysis. Family economy explains trends in school leaving through the early twentieth century, but after that period the pressures of adolescence provides the best mechanism for understanding this phenomenon. Human agency, with students choosing to continue, leave, or resume schooling, overarches all of this, stressing how poor and working-class parents and children proved to be vital and effective historical actors. On the surface, this may appear to be an ambitious, synthetic chapter, but it represents more of a pencil sketch proposing a different way to view the relationship between education and school leaving in urban society. Further, in this chapter, we must draw heavily on generalizations and skip from city to city to suggest the universality of this experience. As David Nasaw (1985) comments about early twentieth-century part-time child laborers,

> working-class Irish kids in Toledo and Philadelphia, Jews in Youngstown and Syracuse, Poles in Chicago and Pittsburgh, Italians in Los Angeles and Cleveland, and 'natives' in cities across the country earned their money after school by scavenging for the same kind of junk, selling afternoon papers, blacking boots, and peddling spearmint gum and chocolate bars to commuters, tourists, and people out for a good time on a Saturday night. Wherever they came from, they were expected to turn in their money to their parents. (p. x)

Social class bound these urban children together, regardless of their location, religion, race, or ethnicity.

Compulsory Schooling

School leaving would be a non-existent problem without compulsory schooling. However, mandatory attendance has assumed many different meanings at various times. The value of schooling, whether perceived by parents or children, sometimes seemed marginal and never remained constant. American historians have too often overlooked this fact. Many poor and working-class parents chose not to send their children to school, but that experience has too often been obfuscated or relegated to minor contextual material. American educational historians, who usually maintain a traditional approach, fixating on the institutional evolution of schools, or a revisionist interpretation, ignoring human agency, habitually overlook parents' and children's perspectives and actions regarding education (Altenbaugh, 1981; Cremin, 1965; Hiner and Hawes, 1985).

J.S. Hurt (1979), a British historian, has taken a different path, and his treatment of English and Welsh parents is instructive for American social and educational historians. Prior to Britain's Elementary Education Act of 1870, which mandated compulsory but not free schooling, laboring parents placed minimal, if not marginal, value on secular knowledge for their children, who would spend their working lives toiling with their hands. These families also saw no need for the school to assume the role of socialization or religious instruction; inculcating values remained the parents' prerogative. For them, schooling simply served the utilitarian function of literacy training, and they appeared to be willing to pay for this service. 'Parents faced a double cost. They had not only to find money for school fees but they also had to forgo the child's earnings. The opportunity cost of losing a child's wages was a far greater burden than mere payment of the weekly fee'. Parents therefore had alternatives, but 'after the decade of the 1870s they lost this freedom of choice'. This dramatically altered working-class culture:

> No longer could parents take for granted the services of their children in the home and their contributions to the family budget. Traditional working-class patterns of behavior, when continued, did so in defiance of the law. The state had interfered with the pattern of family life by coming between parent and child, reducing family income, and imposing new patterns of behavior of both parent and child. (*ibid*, pp. 3, 25, 30–1 and 34)[2]

Hurt (1979) elaborates on the struggles between families and school officials over children in an insightful chapter titled 'Schools, Parents and Children'. During the 1870s, school authorities conducted less-than-legal sweeps through urban neighborhoods 'to round up the children and haul them off to school. In Manchester these sorties into the urban jungle had led to affrays when parents had tried to rescue captives. Police then had to be summoned to rescue the would-be captors'. Many families also objected to the schooling process, particularly corporal punishment. They occasionally dragged a schoolmaster before a magistrate because of excessive abuse. However, few working-class parents challenged the content of what was taught in the classroom, usually acquiescing over religious instruction in the schools. They chose 'not to exercise their legal rights. The "fear of incurring the displeasure of their superiors in the neighborhood and imperiling their employment" was enough to deter them' (*ibid*, pp. 156 and 174).

Human agency culminates in Hurt's (1979) final chapter, 'Unwillingly to School'. Employers and families alike cooperated to evade the law. One factory inspector recalled:

> As soon as an inspector pays one of his rare visits to a 'stocking' village, his coming is at once made known, and children are quickly hidden away, under steps or baskets and in cupboards and holes, and when his back is turned, employers and parents congratulated themselves on their good luck and adroitness in escaping detection. (p. 194)

This collusion included boys and girls, working full-time and part-time and encompassed a variety of industries. One brickfield inspector recounted: 'Report has it that once I stepped on a girl under some matting without discovering her' (*ibid*, p. 196). Seemingly clever inspectors resorted to a variety of ploys in order to surprise their prey, but often to no avail. Unfenced iron mills, for instance, worked against them: 'The boys dropped their tongs and ran like rabbits' (*ibid*).

School attendance, not completion, represented the earliest goal, but attendance remained problematic even well into the twentieth century in Britain. Even as late as 1967, many poor and working-class families placed little economic value on schooling: 'In the working-class home the bright child was the one who demonstrated his (*sic*) academic prowess by leaving as early as legally possible' (*ibid*, p. 211). Thus, as Hurt (*ibid*) concludes about the English and Welsh experience, many — but not all — working-class parents did not support school attendance. Contrary to traditional interpretations and popular notions, a similar mixed pattern existed in the United States.

David Tyack's (1976) comprehensive overview of American compulsory schooling provides periodization and a historiographic foundation. The 'symbolic' stage occurred between 1850 and 1890, when a rudimentary elementary school system emerged which attracted a growing enrollment. 'Most states passed compulsory-attendance legislation during these years, but generally these laws were unenforced and unenforceable' (*ibid*, p. 359). First, many educators preferred not to drag unwilling, usually rowdy and potentially disruptive, students into the classroom. Second, schools, especially those in large cities, simply could not accommodate perfect attendance; these frugal settings lacked both sufficient space and enough benches and desks. Third, 'many citizens regarded compulsion as an unAmerican invasion of parental rights' (*ibid*, p. 361; Burgess, 1976, p. 213; Hawes, 1991, p. 41). Finally, few states enacted enforcement provisions (see Ensign, 1969). During this initial period, therefore, neither educators, the school system, society, nor the state recognized school completion as a desirable — or, for that matter, an achievable — goal.

The 'bureaucratic' phase of compulsory schooling, which began in the 1890s, changed all of this. Tyack (1976) summarizes this stage:

> School systems grew in size and complexity, new techniques of bureaucratic control emerged, ideological conflict over compulsion diminished, strong laws were passed, and school officials developed sophisticated techniques to bring truants into schools. By the 1920s and 1930s increasing numbers of states were requiring youth to attend high school, and by the 1950s secondary-school attendance had become so customary that school leavers were routinely seen as 'dropouts'. (p. 359)

School attendance therefore evolved, ever so gradually, from an informal, almost casual, experience to a formal, seemingly rigid, routine.

Historians explain this change by relying upon five interpretive frameworks to justify the need for compulsory schooling. Those who emphasize political pressures and those who stress ethnocultural factors see the move toward compulsory attendance as a drive for conformity. The former perspective maintains that political forces seek to create a unified nation state, with obedient and loyal citizens, while the latter views compulsion as the means to impose a single tradition and set of values, thus defusing the potential cultural and religious conflicts arising from diversity. A third perspective sees mandatory attendance as the attempt by educational experts to ensure the smooth bureaucratic function of education in a modernizing society. Fourth, human-capital theorists stress how compulsory attendance has contributed to national economic growth. Finally, radical revisionists see compulsion as aiding the reproduction of inequitable social relations (*ibid*, pp. 364–87).

We are not, however, concerned with accounting for the rationale behind compulsory schooling, though we are interested in its impact and permutations. Furthermore, in spite of these different, and often contending, interpretations, this move toward coercion was neither neat nor smooth. Conflict often characterized the issue of school attendance. As Hunt and Clawson (1975) contend:

> As a result of compulsory attendance laws, the schools inherited a problem. They were forced to assume the roles of caretaker and custodian. School became a place a youngster was compelled to attend. Some did not want to go to school. For them, for one reason or another, school was an unpleasant place to be. The schools on the other hand, were legally ordered to keep them. In spite of legislation, even of a primitive nature, and despite the efforts of truant or attendance officers, the situation was 'fluid' and has remained so. (pp. 238–9)

Compulsory schooling represented a state, rather than a federal, initiative, and uniformity simply did not exist in this context: 'With control limited to the states, a great deal of local option and variety remained in education' (Burgess, 1976, p. 205). Certainly every state had passed and enacted a mandatory schooling bill by 1918, but while these school laws required attendance to the age of 14, it was possible in a number of states to avoid attending school. 'Work permits, local options, and a host of other reasons exempted children from compulsory attendance legislation' (Hunt and Clawson, 1975, p. 238). Finally, mandatory attendance, along with other factors, shaped adolescence, which ironically often acted to undermine attendance, as we shall see.

Enforcement of compulsory school laws proved to be difficult. Like their English and Welsh counterparts, many American parents throughout the nineteenth century simply held mandatory attendance in contempt, resenting state meddling in family matters, and often directed their scorn and anger at the person most responsible, the truant officer. Tyack and Berkowitz (1977) quote one indignant Connecticut parent: 'immagane What Sort of person you would be to hold down an Worthless Position for what there is in it. And I only regret that I have to Pay taxes to keep such People that are of no use Whatever' (pp. 32–3). Rigidly moralistic truant officers, on the other hand, viewed the parents of these 'hookey' players as suffering from 'defects of character'; their children were incorrigible. The 'hookey cop' used 'fear and punishment' as the 'appropriate remedies, at least for the hard core'. Boston's school officials used force, intimidation, and punishment to control student absenteeism. Truant officers sent 'convicted truants' to the House of Reformation on Deer Island, thoughtlessly mixing hookey players with true juvenile delinquents. City school systems in general competed with each other 'to report high attendance rates, but to do so they often ignored non-enrolled children and had dropped pupils from the registers if they missed four or five days'. This represented no mean figure, amounting to tens of thousands of children. In New York City alone, during the 1911/12 school year, 'census takers found 22,509 children illegally absent and thousands more out of school because they were ill, neglected, and deserted' (Tyack and Berkowitz, 1977, pp. 36, 41 and 51–2).

Tyack and Berkowitz (*ibid*) argue that a profound change in 'truant work' occurred during the early years of the twentieth century. Reformers and school people tried earnestly to

> attract or push *all* children into school and created elaborate organization machinery to do so. While echoes of the older Victorian moralism persisted, explanations of nonattendance during the Progressive Era focused on broad environmental factors like poverty, illness, and cultural differences. In the 1920s psychologists called attention to the individual truant. (p. 36)

Progressive educators, relying on such 'scientific' approaches, began to see relationships between social institutions like the family, work, religion, the legal system and school as the explanation for truancy.

Child studies pointed to specific causes. In Chicago, for example, researchers studying 'habitual truants' during the 1913/14 school year, discovered 'multiple-problem families': '80 per cent were poor or very poor; 28 per cent had one or both parents dead; about a third were undernourished, and many had serious physical defects or emotional problems' (*ibid*, pp. 42 and 44). Because of these, and other similar, findings of social 'pathologies', the truant officer by the 1920s had evolved into the 'social field worker'. They visited truants' homes, preventing truancy instead of policing it, 'helping rather than punishing children or parents' (*ibid*, pp. 44 and 46).

The school bureaucracy too expanded and changed to fulfill the often elusive ideal of '*universal* education'. New data-gathering techniques appeared to facilitate 'child accounting'.

> University scholars and school administrators developed a complex new technology of enforcement to replace the old haphazard arrangements: a continuing census to find and register every child of school age (often including youth to age eighteen who attended continuation schools while working); hierarchical attendance bureaus which specified roles for supervisors, vocational placement officers, visiting teachers, clerks, and attendance workers; and massive arrays of forms, forms, forms. (*ibid*, pp. 52–3)

School systems also created 'specialized classes or schools' to attract and hold these 'exceptional or rebellious' students, such as 'ungraded classes for "backward" children; disciplinary classes for unruly youngsters; special facilities for handicapped or sick children; and parental schools for hard-core truants' (*ibid*, p. 53).

The relationships among families, children, schooling and work have profoundly changed during the past 150 years, and American historians, like their British counterparts, can capture this experience. What follows is a speculative approach, arguing that the centrality of the family economy through the early twentieth century served as the key to understanding these interactions.

The Family Economy

The ideal of individual choice as a determinant of one's actions and experiences represents a middle-class value, which often did not exist for poor and working-class youths. Walters and O'Connell (1988; see also Hareven, 1982) generalize instead that 'for poorer families, whose economic position was by no means secure, the family is more properly understood as the basic decision-making unit that allocated the time of family members to various activities, including working and going to school' (p. 1124). Historians have too often overlooked this crucial point: 'The culture of the working-class family was guided by collective goals and needs, and the individualistic aspirations of children were subordinated to family demands' (*ibid*).

Women and children often served as the initial manufacturing labor force, operating spinning jennies in Philadelphia and Boston. In 1814, eighty-seven cotton mills nationwide maintained a 'labor force of 500 men and 3500 women and children' (Montgomery, 1968, pp. 12–13). Some early industrialists, as Lawrence Cremin (1980, p. 349) points out, even hired whole families for the emerging factory system. In 1800, Samuel Slater recruited parents and their children for his new and enlarged cotton mill in Pawtucket, Rhode Island. 'The new arrangement relieved Slater and his partners of the responsibilities of oversight in the factories: the parents themselves supervised the children at work and in the process not only maintained social discipline but provided substantial legitimatization for the employment of young children'. This use of the 'so-called family system' not only became widespread, used in Rhode Island, Connecticut and Massachusetts, but persisted through the early twentieth century, as Tamara Hareven argues in her study of the Amoskeag Mills (Hareven, 1982; Horan and Hargis, 1991). Thus, a highly varied, deeply textured, and extremely dynamic working-class structure and culture began to assert itself in an emerging industrial capitalism. And many of those class trends continued through the nineteenth century.

More importantly, working-class families responded in a variety of ways to public schooling, which began in Massachusetts during the 1840s; they struggled to maintain autonomy over their children. Like their English and Welsh counterparts, American families and schools clashed over the fundamental notion of authority. In Carl Kaestle's (1983) balanced study of nineteenth-century educational history, 'school reformers argued for the precedence of state responsibility over traditional parental responsibility for education'. Some parents disagreed, however. 'They took their children's side in cases of school discipline, disrupting school sessions to argue with teachers, in some cases assaulting the teacher and in others having the teacher arrested and brought before a magistrate'. Other parents, protesting school policies, simply kept their children at home; still others subverted school authority, and sowed student defiance by openly criticizing teachers in front of their children. As Kaestle (1983) aptly summarizes it:

> Parents and teachers in the antebellum Northeast were not in a state of declared war, but neither was their relation blissful. The indifference of many parents to school exhibitions, the persistent belligerence of some others resisting school discipline, and the anti-parental propaganda of antebellum educators belie the notion of consensus and collaboration. (pp. 158 and 160)

The 'family labor system', as Thomas Dublin (1979) terms it, appeared to be a universal phenomenon during the early nineteenth century. In the Lowell (Massachusetts), mills, children earned from 51 to 71 per cent of family income, depending on family structure and family cycle. 'Children's earnings must have amounted to more than 80 per cent of family income in female-headed households'. In other families, more children, of course, meant increased family earnings and older children raised family income as they gradually became adult wage earners. This system appeared to be so integrated that 'payroll records for 1850 and 1860 provide repeated instances in which fathers signed for and probably picked up their children's pay envelopes'. School enrollment reflected this reality; 'only 24 per cent of the children attended school'. Gender represented an important distinction, however:

> The vast majority of boys in the 10–13-year-old age group remained in school and did not work, in contrast to girls of the same age. About 74 per cent of boys in this age attended school and did not work, compared to only 52 per cent among girls. Conversely, almost 42 per cent of girls worked, while only 13 per cent of boys of this age did so. Between 14 and 17 the gap in employment rates between boys and girls narrowed until by age 18 the work patterns of the two groups were roughly the same.

Parents believed that schooling possessed more value for boys. Because of a paucity of evidence, Dublin (1980, pp. 171, 173, 174, 178 and 179) presents a speculative explanation, that is, families attempted to extract greater earnings from their soon-to-be-wed daughters. Moreover, families saw some vague connection between schooling and the enhanced earnings of their sons, who would eventually serve as breadwinners.

Kathleen Neils Conzen (1976) found in Milwaukee, between 1840 and 1860, that school attendance depended on social class and ethnicity. Children of German immigrants maintained the lowest attendance level regardless of social position, yet retained the highest literacy rate. In 1850, they comprised 38 per cent of that town's work force, and occupied 'high visibility at all levels of the city's economic ladder', with 20 per cent non-manual, 50 per cent skilled, 5 per cent semiskilled, and 20 per cent unskilled. School attendance, public or private, appeared most problematic for the latter two groups. Children of casual laborers found work to provide immediate relief for their families, particularly if the father had suffered illness or injury. In fact, the whole family worked: 'the mother did laundry, the young boys sawed wood, and the sisters entered domestic service. Other families even more desperate set their younger children to begging or raiding backyard swill barrels for garbage and grease to feed the hogs'. However, some semiskilled and unskilled families managed to apprentice their 12–14-year-old sons as machinists, carpenters, or masons. 'For families who could do without immediate income from their sons, such apprenticeships were a means of ensuring a more secure future for their children' (Conzen, 1976, pp. 59–60, 69, 73, 90 and 91).

A similar strategy guided decisions about Milwaukee's German immigrant daughters. Boys, as in Lowell, attended school more often than girls, who usually found employment as live-in servants. 'Parents were assured that domestic service involved no loss of status, that it meant training for the household the girl would hope to run after her marriage, and a chance to earn a nest egg in the meantime' (*ibid*, p. 92).

Apprenticeship represented a direct relationship between families, children, and work. Cremin (1980, p. 343) describes its colonial antecedents:

It involved a formal contract between a youngster (most often a boy, occasionally a girl), a master craftsman or tradesman, and the youngster's parent(s) or guardian. The most important elements in the contract were the youngster's promise to serve the master in all lawful commands and capacities over a stipulated period of time and the master's promise in turn to teach the youngster the arts and mysteries associated with a particular craft or trade. (see also Rorabaugh, 1986, p. vii)

This implied systematic and comprehensive instruction in the workplace, with the young apprentice residing with his surrogate family. Rorabaugh

(1986, p. vii) adds that while apprenticeship did not fulfill all of its objectives, 'it provided a safe passage from childhood to adulthood in psychological, social, and economic ways for a large number of people over a long period of time'. With republican ideology, which spawned a new sense of freedom and independence, the mechanization of production, which began the long deskilling process, and a persistent labor shortage throughout the nineteenth century, which maintained wage rates, this relationship became less formal, with traditional apprenticeship experiences shrinking from seven to six, or even five, years. Nevertheless, 'apprenticeship remained the most common form of craft training' (Cremin, 1980, p. 344). The Boston school board rationalized low attendance during the early nineteenth century: 'The parents of scholars are able to find places to put them out as apprentices, or in counting houses' (quoted in Tyack and Hansot, 1990, p. 125). This experience was changing, however, as apprenticeship further declined between the Revolution and the Civil War. 'Like a glacier, the institution receded year by year, imperceptibly at first and more rapidly later' (Rorabaugh, 1986, p. vii). Master printers and shoemakers, for example, virtually disappeared by mid-century. Others would soon follow, as we shall see.

Parents reasoned, therefore, that work and work training, not schooling, ensured that they and their children would not only be financially secure but could facilitate occupational mobility. They had direct access to the workplace for jobs and, better yet, apprenticeship experience.

These strategies of choice may be generalized for the entire antebellum period. The common-school movement certainly appeared to increase school enrollment. According to Kaestle's (1983) analysis, a 50 per cent attendance rate at the elementary level represented the northern 'norm' by 1850. 'Indeed, investigation of enrollments for eight towns in Massachusetts, in 1860, for Washtenaw County, Michigan, in 1850, and Chicago in 1860 reveal rates of 85 to 95 per cent at the prime common-school ages, 7–13, for all ethnic and occupational groups'. However, 'pockets' of non-attenders still concerned educators, particularly in 'urban slums and factory tenements'. In Philadelphia's cotton mills 'the working day for children and adults alike ranged from eleven to fourteen hours. One-fifth of the employees were under age 12, and no provision was made for their education. Of all the employees under eighteen, only one-third could read or write'. Employers appeared to be ambivalent concerning child labor; some complied with the weak child labor laws while others painlessly ignored them (Hawes, 1991, p. 41). Some parents likewise violated these child labor restrictions: 'Factory workers needed the extra income' (Kaestle, 1983, pp. 106, 107 and 109).

The emergence of public high schools, seemingly fulfilling the democratic ideal of free schooling for all, again failed to reflect all social classes. 'In the 1830s far more non-farm youths entered apprenticeships than attended high schools that would have fitted them for a wider variety of careers in those segments of the economy that were growing most rapidly' (Rorabaugh, 1986, p. 188). The lure of mechanical pursuits, the reverence for tradition, and the promise of higher earnings continued to attract them. The 1837 depression, according to Rorabaugh (*ibid*, p. 119), marked a change. Many masters went bankrupt, business opportunities decreased, wages fell, and the standard of living declined. High schools now offered the best opportunities. For Kaestle (1983)

> precise studies of high schools in Chicago, New York, and Salem in the 1850s indicate that sons of clerks, merchants, proprietors, craftsmen, and professionals attended. . . . A few factory workers' sons appeared on the rolls, but the lower working class is severely underrepresented. The trend in graduate's careers was toward white-collar work, both clerical and professional, regardless of whether the boys' fathers worked in manual or non-manual jobs. The New York graduates of 1858 included a brass turner's son who became a lawyer, a machinist's son who became a bookkeeper'. (p. 121; see also Tyack and Hansot, 1990, p. 125)

Kaestle (1983) further speculates that the ongoing deskilling of labor and the growing insecurity of work prompted craftsmen, more and more, to send their sons to high school instead of arranging an apprenticeship experience for them: 'Some fathers with such artisan labels may have been substantial craftsmen or even proprietors of their own businesses; however, because of changes from craft to factory production, some members of this artisan group may have felt anxious about their positions and their sons' futures' (p. 121).

The need to work and changes in the work process, therefore, shaped family decisions concerning the importance of schooling. In contrast to an emerging middle class, for children of unskilled parents, the workplace offered them endless opportunities with the growing demand for unskilled labor. Some of these parents needed their children's income to ensure the survival of the family, while other parents saw work as educational, teaching values and skills necessary for economic survival. Yet, ever so gradually, as witnessed by mid-century attendance patterns among some craftsmen's sons, schooling was beginning to function as a bridge to a job. With the decline of crafts, they saw schooling as the source of new values, skills, and job security.

Nevertheless, resistance characterized the relationship of many poor and working-class parents, and their children, to schooling. First, and in general, some parents resisted the imposition of school authority over their children, confronting teachers and administrators about school values and procedures. Second, and in particular, many poor and working-class parents continued to withhold their children from school throughout the nineteenth century. According to 1880 data from the St. Louis schools, '80 per cent of the sons of professional fathers and 64 per cent of the sons of white-collar workers between the ages of 13 and 16 were in school, compared to only 32 per cent of the sons of unskilled laborers'. While many boys chose not to attend high school, girls did, and often in overwhelming numbers: one rough estimate for 1866 indicates that female secondary students outnumbered males by a margin of two to one (Tyack and Hansot, 1990, pp. 131 and 143). Families thus continued to retain control of their children as work grew increasingly unskilled and the struggle for the workplace intensified.

The continuous deskilling of work, which systematically, but not evenly, reduced the need and number of skilled laborers, profoundly affected family income and roles. A 1893 statistical analysis of workers in the coal, iron, and steel industries, for example, revealed that the 'husband's' average earnings only accounted for 85 per cent of the total income necessary to support their families. This report only included the wages of skilled workers, such as 'foremen, miners, engineers, masons, etc.' (Gould, 1893, pp. 17 and 25). They of course received the highest wages, which allowed them to better support their families than their semi-skilled and unskilled counterparts. These latter groups faced serious financial deficiencies. Yet, in spite of the disruptions of industrial life, the working-class family, growing increasingly unskilled and often unemployed, remained intact and stable. According to Virginia Yans-McLaughlin (1971, pp. 302 and 308; see also her 1977 study), the family functioned as a 'working productive unit' in order to resolve the income shortfall of the paternal parent. It employed a variety of strategies, either singularly or in combination, such as taking in boarders, finding employment for mothers, and sending the children to work.

The former two often proved difficult, which left child labor as the only viable option to augment the income of the hard-pressed working-class family. Therefore, in order to survive, poor and working-class families chose, among a variety of strategies, to send their children to work. They often saw little connection between schooling and the workplace: they had access to work and exploited it. Few obstacles existed. 'In 1898, there were still twenty-four states and the District of Columbia without a minimum age requirement for children employed in manufacturing' (Zelizer, 1985, p. 75). With the existence of weak state compulsory attendance laws, like those passed by Massachusetts in 1852 and Connecticut in 1871, which excused poor children from school in order to work, schooling became even more easily expendable (Ensign, 1969, pp. 68–9 and 96–7). Schooling through the nineteenth and early twentieth centuries appeared to be a matter of choice rather than compulsion. Cultural patterns, as we have seen, reinforced this notion, with working children a common characteristic of rural, preindustrial, and early industrial life. This is precisely what Angus and Mirel (1985, p. 139) found in their detailed investigation of the children of textile workers. Focusing on the 1888 to 1890 period, they conclude: 'The decision for a child to enter the work force was exactly that — a decision . . . At any level of father's income, at any occupational level, at any birth position, some children were at school and some at the same age were at work'. Work at a young age was seen as a virtue, not a vice, teaching children diligence, discipline, and responsibility: 'Overwork . . . was a preferable alternative to overcoddling' (Zelizer, 1985, pp. 59, 67 and 100–1; see also Nasaw, 1985, p. 42; Hawes, 1991, p. 41).

Children represented a universal source of labor in industrial America. During a Senate investigation of relations between labor and capital in the 1880s, George Blair, a box manufacturer, testified that children produced between 40 and 50 per cent of the manufactured goods in New York state and a New York City tailor confessed that he saw 6-year-olds working in the nation's largest cotton mill (US Senate, 1885, Vol. I, p. 851, Vol. II, pp. 6 and 67). In 1880, 29 per cent of all males employed in the glass industry were under 16 years of age (US Dept. of Labor, 1916, p. 130). According to the 1910 census, an estimated 2 million children below the age of 16 worked; 26 per cent of cotton textile workers in 1907 claimed to be younger than 16 (Boyer and Morais, 1976, p. 184; US Dept. of Labor, 1916, p. 40; Zelizer, 1985, p. 57). Schooling became so expendable that, in 1914, 60 to 65 per cent of all children abandoned it as early as the fifth or sixth grades (Rippa, 1976, pp. 157 and 159). This trend assumed gender and class differences, however. During the first decade of the twentieth century, '17 per cent more girls than boys completed elementary school'. By the twelfth grade, girls constituted 61 per cent and boys 39 per cent of the student body. Tyack and Hansot (1990) juxtaposed social class and found 'fewer working-class boys than working-class girls continued in school, suggesting that males dropped out more often to work' (p. 171). For Angus and Mirel (1985) class more than any other variable shaped work-force entry: 'Families opting to keep their children in school longer were "better off" in a number of ways: their fathers had higher incomes and higher status jobs, they lived in less crowded homes which they were more likely to own, and they found it possible to live within their means' (p. 139). Working children's high visibility caused one manufacturer to relate an ironic analogy to Helen Todd (1912– 1913, pp. 69–70), a factory inspector, who published her 1909 survey of Chicago's child labor: 'Ever see that box factory in the next block? It's worth seeing. Go into one of those rooms, and you'd think you were in the fourth grade of a Polish school'.[3]

Child workers faced brutal conditions. While the majority of cotton textile workers earned between $5 and $6 per week in 1910, full-time working children younger than 16 seldom made more than $3. Almost half of the children employed in industry worked at least ten hours a day. In the metal trades, 92 per cent of the children under 16 worked more than fifty-four hours a week and 32 per cent of these worked a sixty-hour week (US Dept. of Labor, 1916, pp. 62 and 283; US Dept. of Labor, 1911, p. 28). Schooling did little to increase their earning power, as one young factory worker expressed it to Todd (1912/13, p. 74): 'What ye learn in school ain't no good. Ye git paid as much in the factory if ye never was there. Our boss he never went to school'.

These children learned well their lessons on the job. Like their working parents, they occasionally asserted their control over the workplace, as Nasaw (1985) describes in his treatment of street workers. In 1899, New York City newsboys protested against the distribution procedures implemented by Joseph Pulitzer's *World* and William Randolph Hearst's *Post* by forming a union and organizing a large and successful strike. 'They cemented their informal communities of the street into quasi-formal unions, held mass meetings, elected officers, declared strikes, paraded through the streets shouting their demands, "soaked scabs", and held together as long as they possibly could'. Messengers and bootblacks initiated a sympathy strike, joining the newsies in 'what nearly became a children's general strike'. This militancy spread to other cities, like Rochester, Syracuse, Philadelphia, Pittsburgh, Boston, Cincinnati and Lexington, Kentucky. Nevertheless, while these strikers acted to protect their workplace, 'the New York City union, like most of the other children's unions, was an ephemeral organization with a limited life span'. A second wave of strikes broke out between 1916 and 1918, but ended with mixed results (Nasaw, 1985, pp. 168, 177 and 182–3).

Many children also preferred work because they so disliked schooling; public schools often employed harsh methods (Wehlage *et al*, 1989, p. 31). As Todd (1912/13, pp. 75–6) pointed out in her survey: 'of some 800 children questioned, 269 gave as their one reason for preferring a factory to a school, that they were hit there'. One child described his school experiences: 'They hits ye if ye don't learn, and they hits ye if ye whisper, and they hits ye if ye have string in yer pocket, and they hits ye if yer seat squeaks, and they hits ye if ye don't stan' up in time, and they hits ye if yer late, and they hits ye if ye forget the page'. At times, school discipline proved more oppressive than factory work. 'Nothing that a factory sets them to do', Todd (*ibid*, p. 76) concluded, 'is so hard as learning'.

The cultural milieu aside, early twentieth-century children, like many of their nineteenth-century predecessors, worked primarily because of economic necessity. As Todd (*ibid*, pp. 75–6; see also Angus and Mirel, 1985, p. 139) wrote in 1912, 'a great part of child labor comes from the premature death or disability of the father through industrial accident or disease, or the unemployment of the father through being engaged in an industry which occupies its people only a portion of the year at low wages'. According to one factory owner she interviewed, 'as far as I can make out, the women and children support the entire family'. And children realized the crucial trade-off between school and work, as one related it to Todd: 'Once I worked in a night school in the Settlement, an' in the day school too. Gee I humped myself. I got three cards with "excellent" on 'em. An' they never did me no good. My mother she kept 'em in the Bible, an' they never did her no good, neither. They ain't like a pay envelope'. School offered long-term intangible rewards while work, which remained easily accessible, guaranteed immediate monetary gains and potential future opportunities. A 1916 US Department of Labor (p. 30) study of the conditions of child and women wage-earners determined the value of this trade-off. The report encompassed the cotton and silk textile, clothing, and glass industries and found that children, aged 14 and 15, accounted for 18.3 per cent of their family's incomes.

However, the reign of the street traders in particular and child laborers in general was beginning to wane. The 'high water mark' for child labor occurred in 1920 (Mirel, 1991, p. 1154). Technological, cultural, and educational changes took their toll. 'By the 1920s', according to Nasaw (1985, p. 187; see also Zelizer, 1985, p. 50), 'the children of the city had been pushed to the side by the automobile, which cut off their play and work space, by tougher and better-enforced child labor laws, and by adults who moved into the trades they once monopolized'. Viviana Zelizer (1985, pp. 62 and 63) takes a slightly different stance than Nasaw, moving beyond 'the effect of structural, economic, and technological changes on child labor trends'. A growing unskilled labor supply certainly squeezed children out of the workplace, and rising real income allowed families to keep their children in school. These factors for Zelizer (1985) initiated a gradual change, profoundly transforming 'children's economic roles' (p. 112). Children thus became 'emotional and moral assets' rather than raw economic partners.

And here the collective decision making process within the family asserted itself once again. Between 1880 and 1930, Hogan (1978) found that school enrollment in Chicago increased for all ethnic, racial, and social class groups. In 1930, 97 per cent of the 7–13 age group went to school, 94.6 per cent of the 14–15 age cohort, and 56.6 per cent for the 16–17-year-olds. This represented a changed attitude:

> Different ethnic groups developed distinctive patterns of educational behavior, creating varying matrices of child labor, boarding and lodging, home ownership, and school attendances, in their efforts to ensure survival or enhance their position in the wage labor society. But over time, whatever the initial matrix of educationally-related behavior, all ethnic and population groups kept their children at school for longer and longer periods in order to enhance the economic welfare of their children. It was this positive, instrumental attitude toward education that underlay the increasing levels of school attendance over and above the age of compulsory attendance.

Ethnic culture certainly played a role in these decisions. However, eventually most groups, like Chicago's Slavic immigrants, with the security of home ownership, recognized the 'significance of educational credentials in a wage labor system' (Hogan, 1978, pp. 227, 231 and 255).

Finally, much improved 'compulsory education laws further accelerated the unemployment of children'. All of these factors best explain the ebbing of child labor, since 'effective federal regulation of child labor was only obtained after the Great Depression, first with the National Industrial Recovery Act and in 1938 with the Fair Labor Standards Act, which introduced a section on child labor' (Zelizer, 1985, pp. 63 and 65; see also Hawes, 1991, pp. 52–53).

Adolescence

Mirel (1991) adds that these changes were 'bolstered by the pathbreaking intellectual pursuits of several scholars, who articulated a new school of thought that viewed "adolescence" as a special developmental stage of life, in need of nurture and protection' (p. 1153). G. Stanley Hall's 1904 work, *Adolescence* spearheaded this field of child study. 'While many educators and child advocates were suspicious of Hall's curious blend of science, romanticism, repression, and permissiveness, his basic idea of the "sanctity" of adolescence was tremendously influential'. This drove a movement to 'order . . . the experiences of young people' (*ibid*, p. 1155). For Hall, high schools, functioning as the 'People's High School', should teach practical rather than abstract knowledge, serve a social instead of an intellectual role. He pointed to the industrial schools, prevalent among African- and Native-American students at that time, as his model (Strickland and Burgess, 1965, pp. 22–3 and 149). Schools now began to assume more of the family's responsibilities, thus marginalizing the family as a social institution. An unanticipated outcome was that adolescents now became disconnected from their traditional social institution, the family: 'For adolescents, the most important changes centered on the high school, which was gradually assuming a crucial role in the adolescent life course' (Mirel, 1991, pp. 1155–6).

The actual origins of the idea of adolescence appeared between 1840 and 1880, predating Hall's work. Institutional formalization, according to Kett (1977), changed how youths were perceived. As age-grading evolved, teenagers became identified with the high school years. The concept of development played only a minor role here, however. 'A bureaucratic preference for order and efficiency and the logic of institutional change were the motivating forces' (*ibid*, p. 127). Ability grouping thus fell victim to 'age-segregation'; school authorities sorted out younger and older children at each level simply for the sake of 'systematization'. The term *adolescence*, a generally 'unfamiliar' one, 'began to acquire more specific meanings toward the middle of the nineteenth century' (*ibid*, pp. 3 and 127). The 'teen years' became synonymous with high school.

Changes in nineteenth-century family culture and size also influenced the trend toward adolescence, particularly for urban middle-class families. 'Declining birthrates and the availability of cheap Irish servants cut into some of the traditional functions of teenage girls'. To 'fill up time', they attended high school to become cultured ladies, conforming to the reigning cult of domesticity. 'Possession of an educated daughter became a sort of prestige symbol, a crude form of conspicuous consumption' (*ibid*, p. 138; see also Welter, 1966). In addition to the drop in birthrates, shrinking sibling age ranges and extended longevity reshaped the family and its internal dynamics. This smaller, more intimate, nuclear family unit, for Kett (1977), 'nurtured a more self-conscious approach to the socialization of children; adolescent rearing became as important as child rearing' (p. 232). This 'thrust children into passive and receptive stances rather than active purposive ones' within the family. 'Unproductive dependent' youth, so rare in the early 1800s, appeared somewhat common by 1900 (*ibid*, p. 233).

Finally, intellectual trends and social concerns solidified and formalized adolescence by the turn of the century. Hall's seminal work certainly shaped this movement while child-savers, reformers, parents, and intellectuals drove it. These 'architects of adolescence used biology and psychology . . . to justify the promotion among young people of norms of behavior that were freighted with middle-class values' (*ibid*, p. 243). These norms included conformity, anti-intellectualism, and acquiescence. As a result, 'a biological process of maturation became the basis of the social definition of an entire age group' (*ibid*, p. 215). High schools, and their 'agencies of custody', now produced what Kett (*ibid*, p. 243) calls 'hollow youth'. Nevertheless, these youths found other ways to express and assert themselves, as we shall see.

Within the history of childhood, the twentieth century represents 'the century of adolescence' (Aries, 1962, p. 30). Mirel (1991, p. 1154) divides adolescence into three periods. The first extended from 1900 to 1930, when Progressive educators expanded the high school curriculum to accommodate its new social role. He makes a key point in his analysis of the drive to adolescence: 'These curricular changes were part of the larger process of economic and educational change in which preparation for work eventually supplanted actual engagement in work as the primary vocational experience of adolescents' (*ibid*, p. 1157).

The lines between the workplace and schooling began to blur, thus continuing the slow erosion of collective family culture. Progressive school reformers distorted the clear linkages between economic conditions and attendance. They deemphasized poverty as the cause of school leaving, and instead substituted boredom with schooling. The economic opportunities offered by industrial education, advocates reasoned, would hold students in school. 'Child-labor reform groups and reformist educators played down the poverty motive and stressed the dissatisfaction of the child with his (*sic*) schooling, since to do so strengthened their claims that children did not really work', that is, the notion that dropping out created poverty superseded the idea that poverty caused dropping out (Angus, 1965, pp. 40 and 45). Reformers succeeded, arguing that vocational education provided opportunities to working-class and poor children; they now had something to strive for and would remain in school.

The cooperative education movement signified one of the early formalized expressions of special education for poor and working-class adolescents. With the support of the National Society for the Promotion of Industrial Education (NSPIE), an amalgamation of educators and industrialists, thirteen cooperative schools were developed in eastern industrial centers. In Beverly, Massachusetts, the Union Shoe Company contracted with the local high school in 1909 to train students at its plant for four years. Students worked fifty hours a week for twenty-five weeks at the shoe factory and attended classes at the high school for thirty hours a week during the following twenty-five weeks. While at the high school, however, students continued to participate in industrial training classes for fifteen hours a week (McBride, 1974, p. 211; Violas, 1978, pp. 169–92; Cohen, 1968). Soon training the working class through vocational education programs virtually became a universal experience. Paul Violas (1978, p. 15) notes that, in 1909, 25 per cent of New York's high school population attended a vocational program. The statistics from other cities prove even more dramatic, with 33 per cent in Chicago in 1913, 57 per cent in Cincinnati in 1911 and 56 per cent in Elyria, Ohio, in 1918. The 1917 Smith-Hughes Act legitimized the juxtaposition of schooling with work.

> After World War I public education acquired a new social significance and a new image. Schooling was now conceived as the Great Ladder. Through schooling, social groups formerly caught in the jaws of grinding poverty could lift themselves out of the slums and into respectable working-class status. Working-class children could hope to enter white-collar occupations on the basis of training received at public expense, assuming that they were sufficiently tractable. (Angus, 1965, p. 49)

Vocational education, that is, learning how to work in school, functioned to retain potential school leavers.

Mirel's (1991, p. 1157) second stage, between 1930 and 1950, began with the great depression, which served as 'the turning point in the history of adolescence'. 'By destroying the youth labor market, the depression had closed off virtually every legitimate avenue except the high school through which adolescents could grow to adulthood' (*ibid*, pp. 1158–9). The depression's high rate of unemployment boosted school attendance, with a 64 per cent enrollment rate for 14–17-year-olds.

> Thousands of children went to school simply because there was no place else to go. The schoolroom was warm, if not necessarily friendly, and often the school lunch, perhaps paid for out of the teacher's pocket, was the closest thing to a complete meal the child received. The school became, in many communities, a relief agency; with classes of fifty and sixty in rooms designed to seat thirty-five, it could hardly have been more than this. (Angus, 1965, pp. 65 and 66)

Idleness and hunger, not work, now represented the only alternatives to schooling.

With the depression, the 'youth problem' no longer focused on child labor, but now concentrated on youth unemployment. Mirel and Angus (1985) note a 'profound and permanent shift in the basic relationships between youth, schools, and unemployment'. In order to attract and retain the remaining school leavers, school officials began to shift curricular emphasis. National leaders like Charles M. Prosser, an advocate of secondary vocational preparation and NSPIE executive secretary, supported a 'life-education curriculum in which vocational education played a diminished role'. The rationale for this approach stemmed from the perceived death of entrepreneurship and the emergence of persistent unemployment, with school leavers threatening to become a 'permanent underclass'. Thus, in the Detroit schools, the majority of students now enrolled in courses like 'Personal Service', studying diet, etiquette, and dating. The results appeared dramatic: 'While high school enrollments rose in Detroit by 61 per cent from 1929 to 1939, the number of graduates increased by 237 per cent'. The schools in this case appeared to successfully withdraw students from the labor market. Mirel and Angus (1985), generalizing from this depression-era Detroit school experience, maintain that educators 'shifted the purpose of their institutions away from college and vocational preparation and toward a custodialship of the young based on the conviction that there were not meaningful jobs for them and that their task was to adjust youth to that state of affairs' (pp. 490, 499, 501 and 502).

The Second World War briefly changed this again. Mobilization created high employment opportunities and, as a result, youths once again chose work over schooling. According to Angus (1965), 'the Children's Bureau of the Department of Labor estimated that the number of workers 14 through 17 years of age increased from 1,000,000 in 1940 to nearly 3,000,000 in April, 1944' (p. 87). School leaving rates therefore increased: 'After decades of phenomenal increase, high school enrollments declined by nearly a million from 1940 to 1943' (*ibid*).

However, by the 1950s, the high school became the '"inescapable institution" for American adolescents' (Mirel, pp. 1159–60). Mirel's (1991) third, final and current, stage began in 1950, and is still unfolding. Although the adolescent movement progressed at different rates, more quickly in small cities and towns than in rural and metropolitan areas, by mid-century it appeared to be widely embraced. 'In the three decades after 1920 virtually every state extended the legal protection provided by the juvenile court to those between 16 and 18 or 20. In effect, adolescence became a legal as well as a social category' (Kett, 1977, p. 245). Schools provided education and socialization in their broadest forms, with academic and vocational training, psychological counseling, suicide intervention, substance abuse advising, sex education, and birth control clinics, among others.

All of this was not without conflict, however. 'Even as it became almost impossible to avoid exposure to the institution of adolescence, many young people voted with their feet and resolutely pursued lifestyles that deviated from the idea of conformist and ingenuous adolescence' (*ibid*, p. 246). These youths continued to make choices in the face of convention. Adolescence certainly became identified with high school 'studentship' and 'acquired the status of an official code, but its acceptance did not preclude the emergence of subcultures, or cultures within a culture' (*ibid*, p. 254). Peer group pressures now superseded parental authority and guidance, further undermining the family. 'Adolescents identified with these groups, in part because they saw themselves as marginal, fulfilling no social role other than as consumers and condemned to a "passive dependency imposed on them by schools"' (Mirel, 1991, p. 1161). This subculture adopted distinctive modes of dress, lingo, heroes and heroines, and other symbols, like the automobile, to distinguish themselves. Unlike the collective decision making of the family economy, students now based their actions on peer acceptance or made individual decisions largely divorced from any context. Adolescent culture too often became hostile to schooling. Juvenile delinquency and gangs represented the most extreme expressions of this counterculture. While both manifestations of youth rebellion have always existed, gangs since the 1950s have become predatory and aggressive unlike their more protective and generally defensive nineteenth- and early twentieth-century predecessors. Adolescence, therefore, has not represented a smooth, linear, or monolithic experience, but too often has meant a stormy period of exploration in which young people seek self-identification and social acceptance. Nor have the chronological parameters remained stable. Since the 1930s, adolescent age has steadily dropped, now encompassing 12- and 13-year-olds (Kett, 1977, pp. 265–6). Although juvenile delinquency, which emerged in the 1940s as the new 'youth problem', has increasingly marked adolescence, along with unacceptable homicide and suicide rates, leaving school has emerged as the leading 'adolescent problem' (Mirel, 1991, pp. 1159 and 1162). By the 1960s, as Dorn (1993) affirms, a consensus about school attendance had finally been achieved, sanctifying high school graduation and condemning 'dropping out'; it had become a 'deviant' activity (Wehlage, 1989, pp. 2–3).

Conclusions

School leaving has represented a complex experience in the United States. First, the historical impact of compulsory attendance remains implicit in any analysis of school leaving. Many reluctant students responded like captives, escaping to find entertainment or work, often with their parents' support. Urban schools expanded their bureaucracies to keep records and track these 'hookey players', as well as created 'relevant' programs to attract and motivate such wayward students. Second, in order to understand school leaving in America, we must grasp attendance patterns, namely the interactions between families, children, work, and schools. Throughout the nineteenth century, mere school attendance, not the more ambitious goal of school completion, often created conflict between these social institutions. These relationships profoundly changed, and parents and their school-age children played significant roles in shaping these new connections. Third, with attendance virtually assured by the early twentieth century, school completion became the focus, and adolescence formalized the holding function of public schooling. Adolescent culture too has played a significant role in attending and completing school. Mirel (1991, p. 1163) concisely summarizes this dramatic change and the policy implications for our schools:

> adolescents went economically from being producers to becoming consumers; psychologically, from bearing the burdens of premature adulthood to confronting the problems of prolonged childhood; socially, from easy integration into adult society to an ever-lengthening period of age segregation; and politically, from being the object of campaigns to end child

labor to being the focus of efforts to reduce teenage unemployment. At the center of these changes is the American high school, which itself has evolved from a largely academic and vocational institution into one increasingly concerned with the custodial care of adolescents.

At the end of the twentieth century, poor and working-class adolescents, according to our interviews of Pittsburgh school leavers, face an ambiguous school setting, and naturally question the purpose and value of it. When they cannot readily find these answers, they leave, sometimes unwillingly. More importantly, they often return.

The school leaving rate has steadily, but irregularly, declined throughout this century, yet one pattern continues. From all indicators, poor and working-class teenagers abandon schooling at a higher rate than their middle- and upper-class counterparts, with sharply disproportionate rates among minority students. This trend demonstrates a widening chasm between social classes. However, unlike in the nineteenth century, few alternatives to schooling presently exist. Again, as we found in Pittsburgh, only rarely do school leavers seem able to find meaningful work — or any job for that matter.

Notes

1 Parts of this chapter are taken from Altenbaugh (1993).
2 E.P. Thompson (1967) also treats the impact of industrial capitalism and schooling on working-class culture.
3 Hogan's (1978) award-winning article, led us to Todd's colorful picture of child labor in Chicago.

Chapter 3

The Problem: Context and Approach

... one of the reasons I watched TV. It was time with my dad. We weren't very close. (Pittsburgh school leaver)

Pittsburgh, in many ways, represented a microcosm of the historical pattern of school leaving in America. This city's schools long struggled to educate students, but the lure of work, sometimes meaningful and always necessary in this once heavily industrialized setting, drew them away from school. The recollections of recent Pittsburgh school leavers also suggest that their experiences parallel the lives of their counterparts in other cities; their recollections, however, paint an intricate picture. This chapter summarizes the historical context of school leaving in Pittsburgh, briefly describes our methodological approach and analyzes school leavers' family life and other out-of-school realities and how these factors are related to the school leaving decision.

Attendance Patterns in Pittsburgh

A Historical Perspective

School leaving has plagued Pittsburgh, like other major cities, for decades. As a leading manufacturing center, it offered many opportunities for unskilled workers, adults as well as children. Samuel Jones (1826), an admirer of the town, effusively observed in 1826:

> It is, indeed, not to be denied, that, for almost every kind of mechanical labour and invention, Pittsburgh has acquired a reputation well deserved, and unsurpassed, or unrivalled by any interior town of the United States. The loud strokes of the hammer, and the lumbering wheels are heard within its borders, from the rising to the setting of the sun, and, when often long after. (p. 38)

This was not just hyperbole; Pittsburgh that year, boasting of many iron works and foundries, produced iron, manufactured nails, fabricated steam engines, spun, dressed and wove both cotton and wool, made glass for windows, bottles and tableware, processed paper, milled flour and hosted a variety of other 'manufacturers'. Because of its manufacturing prowess, Pittsburgh earned the nickname of its English counterpart, the 'Birmingham of America' (*ibid*, p. 49).

Ample natural resources and navigable rivers facilitated further industrial development. Pittsburgh's first successful blast furnace appeared in 1859, and within a decade the city became one of the world's leading iron and steel centers; by the late 1890s, Carnegie Steel alone outproduced Great Britain's entire output by 700,000 tons annually. In 1870, its factories produced half of the nation's glass and refined almost all its oil. Industrial and financial magnates Andrew Carnegie, Henry Clay Frick, George Westinghouse and Thomas Mellon built and based their empires in Pittsburgh. Henry J. Heinz, in 1869, launched his food processing business in the city (Livesay, 1975, pp. 165–6; Lorant, 1964, pp. 148, 150, 161 and 232).

But there was another side to the picture. Unfettered economic 'progress' produced serious social problems. 'The work ethic had run amuck: Sunday work, night work, the twelve-hour shift in the mills. The life of the district was keyed to the twelve-hour day' (Lubove, 1976, p. 18). Willard Glazier (1883), visiting the city during the early 1880s, was repelled:

Darkness gives the city and its surroundings a picturesqueness which they wholly lack by daylight. It lies low down in a hollow of encompassing hills, gleaming with a thousand points of light, which are reflected from the rivers, whose waters glimmer, it may be, in the faint moonlight, and catch and reflect the shadows as well. Around the city's edge, and on the sides of the hills which encircle it like a gloomy amphitheater, their outlines rising dark against the sky, through numberless apertures, fiery lights stream forth, looking angrily and fiercely up toward the heavens, while over these settles a pall of smoke. It is as though one had reached the outer edge of the infernal regions, and saw before him the great furnace of Pandemonium with all the lids lifted . . . One pictures, as he beholds it, the tortured spirits writhing in agony, their sinewy limbs convulsed, and the very air oppressive with pain and rage. (p. 22)

Children served as an important component in this wretched work-force, since schooling was unavailable at worst and limited at best.

Pennsylvania state law established public schools in Pittsburgh in 1835, but this early ward system remained decentralized, haphazard, irregular, underfinanced and ill-equipped (McCoy, 1951, pp. 221–2). This severely constricted schooling, since it was simply unavailable for anyone older than 12. Work dominated their lives after this age anyway.

> Children, sometimes very young, were put to work in the glass factories or in the textile mills to work from sunrise to sunset, six days a week . . . When a child was 'bound out' to a tradesman, he was to furnish that child with food, clothing, shelter, and at least three months of schooling each year. On certain occasions, the employer worked the child day and made him attend night school for his education. (*ibid*, p. 223)

'Truancy' therefore beset Pittsburgh's schools, and school officials confronted this dilemma in a most convoluted manner: 'notices were sent to the parents which required their signature and prompt return to the teacher; continued violations resulted in suspension or expulsion from school' (*ibid*).

Attendance remained a persistent problem. African-American students faced complete exclusion from schools until 1837, when the wards created a 'central school for these children'; the state legislature outlawed school segregation in 1881. Furthermore, the state had no compulsory education law until 1895. In addition to work, students found other, extra-school interests, like gangs 'one or more in each ward, and any boy caught away from his own district met with rough treatment at the hands of a rival gang' (*ibid*, pp. 223 and 230).

The schools expanded as the city grew. Between 1835 and 1855, enrollment climbed from 1000 to 3500 students, the number of teachers increased from seven to ninety-seven and buildings more than doubled from four to nine. An 1855 state law mandated a central board, but it actually preserved a decentralized system of school governance. The central board, nevertheless, organized the city's first high school that same year. This modest venture began in a rented building: 'There were two courses open to the students; one for two years for those who could not remain any longer and a regular four-year course' (*ibid*, pp. 224 and 227–8). The city erected its first high school, Central High School, in 1871, with Fifth Avenue and South Side high schools following in 1895 and 1898 respectively. Enrollment spurted from 7416 in 1868 to 46,021 in 1898. This figure jumped to 57,215 twelve years later (*ibid*, pp. 229, 230 and 232).

In 1911, the Pennsylvania legislature passed a School Centralization Act which created the Pittsburgh Board of Public Education. This represented the product of progressive reform fervor, with teachers playing a leading role. The Pittsburgh Teachers Association campaigned for teacher professionalization as well as the enactment of school policies, among them an attempt to curb truancy. This problem was to be addressed through the use of continuation schools (*ibid*, p. 234): 'The Teachers Association wanted special schools to rehabilitate truants. These schools, though costly, would be an asset to the community. Truants could be trained for jobs, and Pittsburgh could be spared criminals' (Issel, 1967, p. 224). Vocational education served as another alternative, enticing students to remain in school by training them for work. The first trade school for boys opened in 1915, the first for girls in 1928 (McCoy, 1951, p. 235).

Teachers' concerns about school absenteeism appeared to be well founded. The 'monumental' Pittsburgh Survey, conducted in 1907 and 1908 and published in six volumes between 1909 and 1914, pointed to child labor as a major social dilemma in that city. 'The most detailed study ever made of an American community, the survey was distinctive for its emphasis upon the effects of the industrial process on human life — the family and home, health and welfare' (Lubove, 1976, p. 18). Margaret Byington's renowned (1910, p. 118) contribution to this study focused on the immigrant working class. All but a few children, between 14 and 21 years of age, worked. In particular, the Slavic families she investigated held school in low esteem; they did not see it as a 'good investment', sending their 14-year-old children instead to work in the glass factories, bowling alleys, or wherever. Parents 'were usually anxious', Byington explained, 'to secure the addition of the children's wages to an income that was truly slender enough'. More than economic reality influenced this decision, however. Slavic parents, Byington (*ibid*) continued, 'felt that what children learned in school had little relation to practical success in life' (p. 160). School knowledge simply appeared to be superfluous in this intensely industrial setting. Beulah Kennard's (1914) contribution to the Pittsburgh Survey condemned, in colorful terms, this 'city of iron whose monster machinery rested neither day nor night' (p. 99). This relentless pace of work and production proved to be pervasive, robbing children of their childhood: 'they literally did not know how to play'. This experience, as we saw in the previous chapter, appeared to be typical.

The new centralized school administration created the Department of Compulsory Attendance, which served as the 'clearinghouse for truants and incorrigibles' (School District of Pittsburgh, 1913, p. 38). It maintained a defensive, nonetheless self-righteous, tone in its 1913 report to the school board:

> Some may not agree with our enforcement of the law, but we believe that the law should be administered with a view to what is best for the child's future, and only those who persist in repeatedly refusing to obey the mandates of the law should feel the full force of it.

This Department claimed thirty-three 'regular attendance officers' and two 'interpreters' in 1913. This staff maintained a daunting pace, making 53,374 visits to students' homes and 1021 visits to employers that year. They filed 133 suits against parents, with ninety-nine resulting in fines or jail (*ibid*, pp. 38 and 39). The Director invoked draconian measures to capture these truants, painting a Dickensian image in his 1914 report:

> We have inaugurated a system of raids on children in the streets during school hours who should be in school. These raids have been conducted in different localities of the city by massing a number of attendance officers and having them patrol the district, arrest all offenders, and bring them before the Director of the Department for a hearing. (School District of Pittsburgh, 1914, p. 86)

The Department's Director blamed parents for this dilemma, but had veiled his accusations in his 1913 report. This time he blatantly reproached them: 'A large number of cases of truancy are chargeable to negligence and indifference on the part of parents. In these modern days the tendency of some parents is to give over to the teacher not only the education of the child but also every other phase of his (*sic*) training' (*ibid*, p. 85). This Director, on behalf of the school district, appeared to be more than willing to assume a custodial role for these truants.

The district's enrollment continued to increase, peaking at 108,000 during the Great Depression: it 'then began to decline'. By 1950, the Pittsburgh schools taught some 80,000 students in ninety-seven elementary and twenty-seven high and trade schools (McCoy, 1951, pp. 236 and 237).

The Current Scene

The Pittsburgh school system presently consists of fifty-three elementary (grades K-5), fourteen middle (grades 6–8), and twelve high (grades 9–12) schools and programs. The district maintains a magnet school program, with a limited number of students selecting specialized academic programs at different sites. During the 1992/93 school year, 40,413 students attended the city's schools; African-Americans comprised 21,266, or 52 per cent of that total. The high schools claimed a total of 11,349 students, with 5427, or 47.8 per cent, African-American.

The district officially defines a *dropout* as 'a pupil who leaves school for any reason except death, before graduation or completion of a program of studies without transferring to another school or educational program'. This district therefore excludes from its school leaver figures any school-age student who enters federal programs like the Job Corps and correctional institutions, or those who return to school. It collects and analyzes school leaver data based on four-year cumulative and annual percentages. The cumulative school leaver rate since 1983 has averaged 24.7 per cent, ranging from 28.4 to 18.2 per cent. The annual rates have averaged 6.9 per cent, with a low of 6.0 per cent in 1987/88 and a high of 7.6 in 1988/89. Sharp disparities also exist between buildings. The annual school leaver rates for high schools during the 1992/93 school year ranged from 1.1 per cent at Perry Traditional Academy to 16.0 per cent at Letsche Education Center.[1] This can be misleading, however. Some highly rigorous programs, like that at Perry, can transfer students back to their home institutions or to alternative schools, like Letsche, to mask their school leaver rates.

Pittsburgh students typically abandon school sometime during the tenth grade at age 17. Race plays a major role since African-Americans leave school at a higher rate than any other group. With a district-wide 7.0 per cent school leaver rate in 1992/93, African-American males and females left school at a rate of 9.6 and 7.3 per cent, respectively. 'Rates for other' male and female school leavers, meanwhile, were 7.0 and 4.5 per cent.[2] How did schooling look through their eyes?

Data Collection

Our goal in this study is to understand how school leavers perceive schooling, to reconstruct it as they saw it, to describe their realities and culture. Spradley (1979) operating within a symbolic interactionist framework, defines culture as '*the acquired knowledge that people use to interpret experience and generate social behavior'* (p. 5). Schools, on the surface, appear to be monolithic, but, upon deeper analysis, host a myriad of cultures. As he states:

> Our schools have their own cultural systems and even within the same institution people see things differently. Consider the language, values, clothing styles, and activities of high school students in contrast to the high school teachers and staff. The difference in their cultures is striking, yet often ignored. (*ibid*, p. 12)

Each group creates its cultural meanings and symbols. Within this dynamic setting, we attempted to isolate school leavers' perceptions of schooling and gain insights into their culture, by using an ethnographic approach, relying on interviews of the participants.

Our research 'adventure' proved to be neither linear nor uneventful. We certainly began with intellectual curiosity, forming scholarly questions; this led to the adoption of an appropriate and fruitful research approach. Executing it and locating informants proved to be difficult and discouraging at times, however. We experienced many false starts and occasionally found ourselves in blind alleys.

Preparation for research began in 1985, and encompassed several meetings, dissecting the interviewer's role, creating the initial set of questions, and assembling a sample of informants. We dwelled on interview mechanics, and a concern to minimize the distorting effects of bias seemed to dominate many of our early discussions. We became aware of and sensitive to warnings, like William Cutler's (1971, p. 2), concerning the interviewer's complex and delicate position:

> Interviewers should always know their biases and conceal them as far as possible. Prejudicial remarks, emphatic intonations, or even a simple affirmation, if consistently applied, can distort a respondent's account of the past. Interviewers themselves are vulnerable to error because of their own biases since a strong point of view, if unrecognized, can induce a narrow plan of questioning accidently tailored to suit private principles and assumptions. Those same prejudices can also cause an interviewer to make hasty judgements about respondents, branding them with a stereotype which can then wrongly guide the rest of the interview. (p. 2)

We too believed that rapport with the narrator represented a critical component of a successful interview, and a 'safe' environment seemed to be a priority. We therefore agreed to conduct interviews at the informants' homes or in other settings chosen by them (Davis, 1977; Seidman, 1991). Finally, and complying with Spradley's (1979, p. 25) scheme, we saw informants acting as teachers for us, the ethnographers.

We prepared a list of largely open-ended questions (see appendix) in order to reconstruct schooling through their eyes and recapture the factors affecting school abandonment as well as resumption. We stressed in-school variables, but also pursued out-of-school influences. These were 'probed systematically' through cues to ensure the uniformity of the information collected. This approach facilitated the comparison of separate experiences and perceptions, and served to strengthen potential conclusions (Bodnar, 1982, p. 3; see also Dohrenwand and Richardson, 1964; Oblinger, 1978; Seidman, 1991; Spradley, 1979).

Our interview guide covered four specific areas. The first part, which also operated as an interview warmup, focused on the student's background, namely age, residence, family size and structure and the student's family responsibilities. The second section concentrated on school factors, like teachers, administrators, peers, academic subjects, extra-curricular activities and the school's physical plant. The third component pursued students' extra-school experiences, examining social interactions, work demands, racial discrimination, neighborhood environments and geographical mobility. We completed this instrument by inserting some follow-up questions; this involved the repetition of some questions to encourage reflection. We reviewed, adjusted and refined the protocol during early interviews. Circumstances prevented us from conducting the ideal three interviews, i.e., 'focused life history', 'details of the experience', and 'reflection on the meaning' (Seidman, 1991, pp. 11–12). Our three-part guide compressed these three important areas. We thus attempted to extricate the best information within extremely tight restrictions.

Locating informants proved to be a daunting task, and serendipity played a role. As Hammarberg (1971) asserts, the sample of narrators must somehow be 'a universe in miniature', in spite of 'incomplete enumeration' (pp. 542 and 561). We fruitlessly pursued the school leavers listed on computer printouts given to us by a high school counselor. Except for one name, the addresses were outdated; we quickly realized that we had encountered a seemingly transient population. We turned next to community organizations and agencies. Our efforts here too failed. We then turned to the Pittsburgh Job Corps Center. This resulted in a major breakthrough, and it also profoundly altered the scope of our project. We discovered that the school leavers who enrolled at Job Corps had decided to resume their schooling by participating in a reentry program. Our focus now became two-dimensional: it dealt with dropouts as well as with a group that has been seldom studied and analyzed, the *dropbacks*.

The Teledyne Economic Development Company has managed the Pittsburgh Job Corps Center, along with several other centers, since the program's inception in 1965. Local state employment offices and social agencies provide information about the program. Students apply, and after a screening process, they are assigned to a center. The Pittsburgh Center serves students from Pittsburgh, Philadelphia and Erie, and as well as rural Pennsylvania, Ohio and West Virginia.

We met with the Center's Director, and formally submitted our proposal. After its approval, staff members arranged for us to interview batches of Pittsburgh school leavers. However, since the flow of Pittsburgh school leavers through the center appeared to be uneven, our interview schedule followed an equally irregular pattern, mixing bouts of frustrating inactivity with short, intense interview sessions.

We interviewed 100 informants between 1986 and 1994. We established this ambitious interview goal, but remained open, willing to terminate the interview process at a lower number or extend it beyond that initial target. As we approached this total, we realized that it proved to be realistic; we had reached the 'saturation' point, with a great deal of redundancy. Seidman (1991) is explicit about such a situation:

> The method of in-depth, phenomenological interviewing applied to a sample of participants who all experience similar structural and social conditions gives enormous power to the stories of a relatively few participants. Researchers can figure out ahead of time the range of sites and people that they would like to sample and set a goal for a certain number of participants in the study. At some point, however, the interviewer may recognize that he or she is not learning anything decidedly new and that the process of interviewing itself is becoming laborious rather than pleasurable. That is a time to say 'enough'. (p. 45)

We have protected the identity of all informants by maintaining their anonymity.[3] We told them about our objectives and took safeguards to preserve their rights and privacy. All narrators were volunteers, signed consent forms, and were told they could terminate the interview process at any time or ask the interviewer to turn off the tape recorder (Seidman, 1991; Spradley, 1979). A few interviews produced extremely sensitive material, and we purposefully destroyed those tapes,

Table 3.1: *Profile of the Interview Sample, by Gender, Race and School Affiliation*

School Building	Males		Females		Dropout Rate per School 1992–93[1]
	African-American	White	African-American	White	
Allderdice	2	2	1	1	4.1%
Brashear	4	1	4	3	7.4%
Carrick	0	0	2	4	7.1%
Langley	2	0	4	0	9.7%
Letsche	4	1	13	1	16.0%
Oliver	6	0	6	0	6.8%
Peabody	5	1	5	1	6.0%
Perry	0	0	2	0	1.1%
Schenley	0	0	1	0	5.1%
South	1	0	0	0	1.6%
Westinghouse	11	0	9	0	9.2%
Arsenal M.S.[2]	0	0	1	0	—
Reizenstein M.S.[2]	1	0	1	0	—
Total	36	5	49	10	6.7%

1 Figures are from 'Pittsburgh Public Schools 1992–93 Dropout Report', Division of Student Information Management, December 1993.
2 Dropout figures were not given for middle schools.

necessitating a search for yet more school leavers. The three of us often met and debriefed each other after interview sessions. We had most of the tapes transcribed. We all read the typescripts, indexed them and met periodically to compare our interpretations and perceptions.

Rather than describe what narrators said, in the following chapters we will allow them, as much as possible, to speak for themselves and for other school leavers. We have preserved their language, and speech patterns, because, as Spradley (1979) stresses, language is 'a tool for constructing reality' (p. 17). We attempt to understand and analyze dropback recollections within the context of school leaver literature as well as historical experience, social theory, and philosophical principles.

Informants

Most of our narrators, in many ways, fit the stereotype. They were predominantly minority students — strictly African American in that sense — and came largely from non-traditional households. We only interviewed two immigrants: one from Italy and another from Trinidad. Table 3.1 illustrates the gender and racial backgrounds of our informants, as well as the high school they attended. Every Pittsburgh high school is represented and we have provided each building's annual 'dropout' per cent for comparison. Females represented 59 per cent of our 100 dropbacks, exceeding the district's figure. African Americans comprised 85 per cent of our interviewees, also outstripping the proportion of African Americans in the district schools. African-American women alone constituted 49 per cent of our sample. Ages ranged from 16 to 23 years. The mean age was 18.56 years, and the mode was 19. Their ages indicated that their school experiences were very recent. Some had indicated that they had abandoned school only weeks before; most were within months of having left. A very few, especially those in their twenties, had been out of school for a few years. We also used information from biographical statements to shed some light on their social class backgrounds and family structures and relationships.

Social Class Background

Social class, as we have seen, has long represented a key variable defining school leaving patterns. Little has changed. During the past fifteen years, and regardless of race and location, socioeconomic background has been directly correlated to school leaver rates:

> Among youth in the upper-upper class, only 2 per cent drop out, while at the lower-upper class the figure rises to 10 per cent. In the upper-middle class 17 per cent of the youth do not complete school, and at the lower-middle and upper-lower class, the rate of school dropouts rises to an incredible 25 per cent. At the lower-lower class, a full 50 per cent of the youth quit school. (Beck and Muia, 1980, p. 67)

During the 1992/93 school year, 38.6 per cent of the student population in the Pittsburgh schools received public assistance. The district, using information from the Department of Public Welfare, decides which students are eligible for free and reduced price lunches based on family income. While 62.3 per cent qualified for this benefit that year, only 25.1 per cent overall actually applied for it, and only 16 per cent of high school students used the free and reduced lunch option. According to school district data more students have been applying for this benefit each year.

Judging from their parents' or guardians' occupations, as domestics, nurses, construction workers, security guards, or their unemployed status, and from the family's place of residence, in the Hill District, Homewood, or Garfield sections of Pittsburgh, we concluded that our narrators came mainly from poor, or at best working-class, families and communities, primarily African-American sections of the city. To make matters worse, 50 per cent of our informants indicated that their parents or guardians were unemployed or collecting welfare or a pension. We failed to determine whether those in the employed cohort were working full-time; many of these were likely underemployed, either laboring at part-time jobs or finding irregular employment, such as construction work. Moreover, even full-time work is often poorly paid.

These school leavers came from economically distressed families. This profile fits with national patterns and reinforces the correlation with minority status: 'In 1987, about 40 per cent of Black and Hispanic children were living in families with incomes below the poverty level, compared to 25 per cent for White children' (Rumberger, 1991, p. 73).

Family Relations and Structure

All of our informants appeared to be disconnected from major social institutions, and this began with their families. According to the literature, families of school leavers fell into two categories: one in which the family hinders the student's chances of school success, and another in which the family meets the student's emotional and physical needs and provides security. In the latter case, the school is solely responsible for student failure, disengagement, and abandonment. We discovered that our narrators' families, while certainly fitting these two extremes, also pointed to a broader experience that, somewhat refuted these simplistic analyses. When informants discussed family relationships, only sixteen reported having outright conflicts with family members. One proclaimed himself to be a runaway. A female African-American narrator left home because of disagreements with her mother: 'We get along if we are sisters. But don't try to be my mom. She had a bad attitude and it don't mix'. Another female dropback related a similar, but more complex, experience. She, at first, claimed that she left school because of problems at home. However, as the interview continued, she elaborated how these domestic conflicts caused her to be angry, and this emotional turmoil understandably spilled over into the school. When this occurred school administrators suspended her. They punished her for the symptom, but overlooked the cause. She eventually just quit out of frustration; a decision she later regretted. Finally, a White male blamed his family situation for abandoning school: 'School wasn't the problem . . . family problems . . . not getting along with my stepdad'.

The experiences our informants described generally fit those outlined in other school leaver studies. Beck and Muia (1980) in their literature review, cite a study in which 62 per cent of school leavers indicated that 'their home lives are unhappy' (p. 66). Nevertheless, these narrators typically responded as one male student did. He watched television eight hours a day, and described how his relationship with his father centered on that activity:

> I liked TV. My father sat there and watched *All in the Family* all the time. He never missed an episode and I would sit there with him. It was the few times I got to see my father. That was the time I spent with him sitting down and watching TV. And I learned to crack at commercials . . . laugh and joke and make some comment about the commercials. That's how we talked. I guess that was one of the reasons I watched TV. It was time with my dad. We weren't very close.

Beck and Muia (*ibid*) summarize the same basic experiences: 'The dropout's family is less solid, less informed by a father figure, less likely to interact in leisure activities, and less able to communicate than the graduate's family' (see also Conrath, 1984, p. 36).

However, we also found that family life was not necessarily a direct cause for leaving school, much as Farrell (1990) discovered in interviews with at-risk students in a New York City high school: 'The data do not lead me to believe, for the most part, that adolescents drop out of school because of family friction' (p. 66). Eighty-four of our narrators described, or at least implied, having cordial relationships with their families. However, this does not preclude the eruption of family crises, which could compound matters. Diedre Kelly (1993), as well as other researchers, finds that girls experience more family demands than boys: 'Family problems included divorce, alcohol and drug dependency, incest, financial difficulties, death of a parent (because of illness, accident, drug overdose, or suicide), and physical and verbal abuse' (p. 158). Some of our female informants reported similar difficulties. At the same time, a male narrator had to cope with his seriously ill mother.

Our narrators consistently described non-traditional family structures. Seventy-six per cent either lived with one parent, usually their mother, and siblings, or lived with a guardian, generally a female relative like a sister, grandmother, or an aunt. One female student 'moved from different homes at least five times'. She lived with her sisters, mother, and godmother, who 'was on drugs'. Ten of the dropbacks claimed that they lived alone. Only fourteen informants described a traditional family structure. Occasionally the household of our female narrators included their own children. An African-American female dropback, when asked what she liked to do for 'fun' expressed tenderness and affection for her son: 'Take care of my baby; watching him grow; the new things he learns and says every day; things he sees that I don't see; just taking care of him; taking him places; teaching him different things; read books to him; let him watch *Sesame Street*'.

After carefully analyzing the interview typescripts, we discovered complex and variegated realities. Indeed, the overwhelming majority of our narrators described non-traditional family settings, but this did not necessarily imply dysfunctional families, just alternative structures. One African-American female, who described a single-parent, female-headed family, had four brothers and sisters. All but she had earned a high-school diploma or GED; at the time of the interview, she was working diligently to earn it, and appeared proud of that accomplishment. This non-traditional family structure did not seem to be dysfunctional in any way. In other cases, these young people described caring and nurturing families. Another female student, who said she watched up to eight hours of television per day, recalled a strict grandmother: 'I watched a lot of TV. 'Cause, after you do homework, my grandmother wouldn't let us go outside, so I would watch TV besides looking at the four walls'. Her grandmother, aware of the dangers lurking in that neighborhood, protected this child by safely confining her to the house. Vicarious experience, via television, represented the only escape in this siege-like environment.

Conclusions

Poor and working-class children have seldom fared well in Pittsburgh's schools. Fitting into the larger historical context, they had to work to support their families and themselves. Over the years, the schools resorted to a variety of strategies, some of them harsh, to ensure their attendance.

Our informants, mostly female and overwhelmingly African-American, also describe distressed economic backgrounds. They pointed to unemployed parents and guardians who are struggling to exist. Non-traditional family structures compounded this situation. These narrators painted a picture of highly stressed families — that is, families in which a single parent, or guardian, is attempting to survive economically while nurturing one or more children in hostile, if not outright dangerous, neighborhoods. The family, the most significant and fundamental social institution for our informants, generally appeared to be in crisis, unable, but not necessarily unwilling, to help them. We therefore conclude that a non-traditional family structure does not in itself cause students to leave school. Rather, one less adult is around to help and care for the child. At best, the family described by our students is a strained institution, unable to offer complete assistance; thus, while it may not cause a child to leave school, it also may not totally mitigate that experience (Wehlage *et al*, 1989, p. 51). All of the social institutions that these narrators came into contact with fell short in some way (see Rumberger, 1991). The school, the second most important social institution for these narrators, rarely addressed their vital educational and social needs. These factors had a cumulative effect. This process will become clearer as we describe and analyze our findings on school settings and personnel, the students' academic life and work, and the actual experience of abandoning school.

Notes

1 Letsche school maintains grades 8–12, unlike other high schools which include grades 9–12.
2 All of the data in this chapter have been culled from 'Pittsburgh Public Schools 1992/93 Lunch Application Analysis Report' (Division of Student Information Management, 5 April 1993), and 'Pittsburgh Public Schools 1992/93 Dropout Report' (Division of Student Information Management, December 1993).
3 We conformed to the Human Subjects Guidelines required by the University of Pittsburgh. We also received clearance from the Department of Labor through the Pittsburgh Job Corps Center.

Part II

The Pittsburgh Study

Chapter 4

The School Setting

People could walk in there with anything.
I believe you have more advantages of being White,
cause this is a White man's world. (Pittsburgh school leavers)

The Physical Plant

The concomitant growth of school size and expansion of many bureaucratic layers represents a general historical trend for the nation as
well as Pittsburgh (Altenbaugh, 1992; Bennett and LeCompte, 1990,
p. 252). While some proponents have pointed to cost effectiveness,
others have boasted about the expansion of curricular choices and extracurricular programs which consolidation and scale produced. Critics
argue differently: 'in schooling, big is not necessarily better. Increased
size has come at the expense of the sense of community and belonging
which smaller schools often could create' (Bennett and LeCompte, 1990,
p. 252).

Too many of our informants felt overwhelmed by the sheer size
of their high school buildings. An African-American female narrator
recalled, 'It was too large. Half the time that was why I was late for
class. I couldn't find my way to class half the time. It was too big'. A
White female interviewee, reinforced this perception when she described
another school building: 'It was two separate buildings connected. It was
easy to get lost'. She claimed to be lost several times. Another African-
American female informant compared her experiences in large high
school buildings with small ones.

> I like small classes . . . There was just too many people. Even
> my social workers know that I can't operate in a school where
> there is a lot of people . . . I like the small schools the best. But
> all the schools I went to was big except Letsche. I didn't start
> going downhill until I got into high school.

All of the students complained about excessive noise, particularly between classes, and few saw the school's physical environment as conducive to learning: 'When you cut classes as much as I did, you got to see a lot of people in the halls. Quite a few of them. No one cut gym' (laughs). One student complained: 'Yeah, it bothers me. Like in class, you know, you trying to study and everything. Be too noisy; you can't concentrate'. Larger schools contribute to this problem.

The rationale for larger schools stems from the belief that 'bigger is better', that is, larger schools, because they are able to concentrate resources, offer a more diverse curriculum and extracurriculum. Students, as a result, have access to more opportunities. Pittman and Haughwout (1987) confront this conventional notion. First, they point out, scale does not simply translate into advantages. 'On the average a 100 per cent increase in enrollment yields only a 17 per cent increase in variety of offerings' (p. 337). Second, smaller, student bodies ensure better access to teachers, activities, and school services: 'students who attend small schools participate in more activities and receive a greater diversity of experience' (p. 338). Third, students who attend large schools demonstrate little school 'identity' or 'commitment'. Large school settings, Pittman and Haughwout conclude, contribute to 'reduced individual participation in school activities, decreased attendance, and less expressed satisfaction with school' (p. 338). Students in such settings often do not experience 'social integration' and tend not to engage schooling (p. 343).

More important, a correlation exists between school size and school leaving, particularly for minority students (Bennett and LeCompte, 1990, p. 252). Stoughton and Grady (1978, p. 314), in their Arizona study, found the following startling data: 'The highest dropout rate occurred in schools with student enrollments of 1000–1500 students . . . In small schools with up to 200 enrollment the dropout rate was half the overall annual rate in Arizona'. Small school buildings in their investigation remained flexible, fostered a more personal relationship between teachers and students, and appeared more committed to nurturing students. Pittsburgh's eleven high schools maintained an average enrollment of 946 students during the 1992/93 school year. Size ranged from 303 students at Letsche to 1436 at Allderdice, with six claiming over 1000 students each year.[1]

Pittsburgh relies on large high school buildings, but it keeps its class size small, averaging 18.7 students. This average encompasses several categories which reflect a broad spectrum of subject areas. Academic courses, like English, social studies, etc., host twenty-one students each, while non-academic classes, such as music and physical education, typically enroll 24.2 students. Vocational subjects, i.e., business education and home economics, accommodate 14.5 students, but gifted and special education courses maintain the smallest number, with 13.9 and 6.8 students, respectively. In spite of this modest average class size, our informants claimed that they attended large classes. Why did they perceive this? As we will see in chapter 5, traditional teacher pedagogy, with desks lined up in rows, deemphasized individual and small group learning. Our narrators did not like this impersonal approach. It further interfered with their recollections of relatively small class size, that is, instructors implemented a teacher-centered, whole-class fashion leaving students with the impression of large classes; teachers simply did not take advantage of small classes to alter their pedagogy.[2]

The Social Setting

Youth culture defines the high school social setting. Bennett and LeCompte (1990, p. 75) treat it along seemingly traditional lines:

> Culture is the set of distinctive practices and beliefs, while the peer group is the social entity which develops and carries it out. To the extent that the youthful peer group is defined by its relationship to school, we can speak of a student peer group. While peer group influence is important in the lower grades, it becomes more profoundly problematic for schools during adolescence.

This culture encompasses four types of behavior: conformity, attempting to appear the same as one's peers; rebellion, contesting adult authority; idealism, embracing a simplistic sense of right and wrong; and immortality, taking risks like using and dealing in drugs, engaging in unprotected sex, and skipping school (*ibid*, pp. 75–6). Adolescence nurtures youth culture, because

> the longer a society schools its children, the greater will be the tendency for young children and adolescents to develop a youth culture which is in opposition to adults, even where students change schools frequently. This is because it is only in age-graded schools that so many people of the same age and sex are forced by law to remain in close proximity for so long a time. (*ibid*, p. 80)

As we have seen, this has been a persistent historical trend in American society.

This dynamic setting produces winners and losers. Successful students accept, or negotiate with, school values, routines, and authority. Unsuccessful adolescents often resist schooling and all of its meanings. Bennett and LeCompte maintain a critical perspective by stressing human agency.

> Resistance is principled, conscious and ideological. The non-conformity of resistors has its basis in philosophical differences between the individual and the institution. It involves 'withholding assent' from school authorities; students who resist may disagree with the way they are treated in the school on the basis of their gender, ethnicity, or social class; they may disagree with the academic track or category to which their gender, ethnicity, academic or social category is held. They become resistors when their disagreement is actively expressed. (*ibid*, p. 106)

Opposition accordingly is not blind misbehavior.

Students exercise this counteraction in many ways. 'Boredom and alienation' mark the 'initial phase'. Students cease studying and pursue activities, like working, raising a baby, or participating in illicit enterprises. 'Assertiveness' occurs when youths confront school authorities and common notions of academic success, maintaining strong and different opinions and rejecting adult notions of conformity. Students may also pursue 'hyper-achievement' in non-academic pursuits as an alternative route to status, especially for those who experience academic failure; some of these may be acceptable, if not laudable, like sports, while others may be illegal. 'Tuning out', yet another approach, actually involves dropping out while still enrolled in school, such as sleeping through classes or ditching them altogether. In the latter case, 'students spend more time in the hallways and hiding places of the school than in class'. This often overlaps into the final manifestation: leaving school, the 'ultimate form of resistance' (*ibid*, pp. 107–9).

Adolescent conformity, as part of youth culture, often acts to reinforce the decision to leave school. Beck and Muia (1980, p. 68) describe two possible scenarios. In the first case, 'potential dropouts' form 'social contacts' with similar peers: 'many of those persons selected for reinforcement are likely to be dropouts or in the process of dropping out'. These peers only act to 'validate' or 'reinforce' disengagement from schooling. 'Contact with these persons maximizes the probability that this dissatisfied adolescent will choose to quit school as a "solution" for feelings of insecurity and failure'.

One of our male African-American informants described this very experience. He and his friends used to 'cut classes, get high on pot, breaking into lockers'. When asked why he did this, he responded: 'I didn't, I was just there. I was with the wrong crowd. I got kicked out'. A White male student painted a similar picture, hanging 'around in the halls, swearing'. He and his friends were suspended for these activities. A female African-American student paralleled these experiences: 'we went out drinking, running with boys, just cutting school. We would go downtown and just walk around'. Substance abuse for her involved 'beer' and 'marijuana'. She too was suspended.

In the second case, when potential school leavers associate with more successful students, their shortcomings become exaggerated, that is, their fear of failure and rejection feeds their school leaving decision.

> If, because of economic or parental constrictions, this individual is unable to meet his/her friends' demands for conformity in leisure activities, ways of dressing, and materialistic possessions, he/she may opt to end the problem by leaving school. The pressure to conform is often so great that failure to do so confirms the individual's self-concept of inadequacy. (*ibid*)

Their attitudes, therefore, may be summarized in one word: estrangement. They express 'feelings of alienation from their schools, homes, neighborhoods, and/or society in general'. Repeating a grade, and concomitant separation from agemates, only compounds these feelings. They seldom, if ever, participate in extracurricular activities. 'Thus, to them, school is nothing more than an overdemanding, unfriendly environment, in which they are destined to fail. It is little wonder, then, that so many choose to escape from this unpleasant situation by dropping out' (*ibid*).

We too saw this experience through one informant's eyes. A female student described social peer pressures, her fights, then withdrawal from the social scene and ultimately from school:

> Some kids might not have what other kids have, material things, something that can be burned. And most kids like to brag and boast about it. Maybe hair — I got more hair than you. So, that makes you a less person, right? And some kids are very weak hearted. Some kids can be very emotional and me, I have to admit, I'm a very emotional person. I have my ups and I

have my downs. It makes you shut out from all the rest of the school. And it seems like, you get the point and an attitude, I'm not going to come there. You know why? 'Cause the kids there get on my nerves. They make you want to cry or something. And then you never have a thought about the teachers or your education. Like me, I just stayed home. My dad asked me why don't you go to school? I don't want to go. He would tell me to get up and go to school.

Her father's cajoling proved unsuccessful.

Some narrators did not fit so neatly into these patterns. On the one hand, they seldom claimed any emotional attachment to their peers, neighborhoods, or schools. These students recalled changing residences and schools a great deal. One narrator noted that she had attended six separate school buildings, excluding elementary schools; a male student counted eight secondary buildings. Students indeed appeared to be mobile. Of the total student population of 43,727 during the 1992/93 school year, 17.7 per cent transferred schools at least once. Most changed schools once, but many switched four or more times. Secondary school students maintained the highest transfer rate, with 20.9 per cent.[3] They certainly felt disconnected from the social scene at their schools. On the other hand, while most of our narrators claimed to have friends, forty-seven stated they preferred to be alone. They appeared reflective. One student did this while fishing: 'I liked to go fishing. I go to Neville Island . . . I ain't really good at it. It's fun though. A lot of people don't like to sit there and just be bored. Just waiting for a fish, it is all right to me'. This penchant for thinking created school problems for another student:

I found out that I like walking. I found out all about Squirrel Hill and Point Breeze . . . I like walking. I loved the scenery. I loved the houses. It would give me time to think. Those walks took long and I missed a lot of school. At the time (in middle school), I came into a pattern of being by myself. I enjoy thinking . . . singing and writing.

Still another student summarized his attitude in this way: 'I don't have too many friends. I don't like loud people. I don't like loud people.'

School leaving, of course, completely severed already tenuous social ties with peers. Few informants spoke about maintaining any contact with their former classmates. A few viewed their abandonment act with regret because of the complete social ostracism they experienced, as in the case of one African-American male narrator. Although he wished his former classmates well face-to-face, he expressed envy to the interviewer: 'I'm like real jealous and everything. I seen a couple of my friends, like a month ago, and congratulated them. They asked me to come to the thing before the prom'. He did not attend any of the prom celebration.

A Personal Perspective

A White female informant found refuge with neither adults nor peers. With her parents divorced, she experienced no solid home life: 'I had a mother that didn't take care of me, and she threw me to my dad. So I bounced'. Adults at her school appeared to be calloused as well. One counselor acted in an insensitive manner. 'I got called into the counselor's office for beating up this kid for pulling my chair out from underneath me. My mother came in and beat the shit out of me in that office and (the counselor) should have at least thought to call someone, but he didn't do that'. The counselor compounded this scene by suspending her from school for fighting. Teachers too seemed to be unfeeling:

> I just don't think that they cared, because I can refer to one instance. This one teacher was having a problem with one of the students. He said (to her), 'if you don't like it you can just get up and leave. I don't care'. So, she got up and left. 'Does anyone else want to go?' So half the class just got up and left.

Adult contacts proved to be indifferent at best and hostile at worst.

This narrator described a complex social scene. She appeared to be a loner in school; she simply didn't get along with her peers, ''cause they had their little cliques'. Students had also stolen her bus pass, forcing her to walk one hour each way to school. This interviewee decided that other social outcasts represented her only haven. She abandoned school at the end of ninth grade, at age 15, and married. She began to associate with the 'wrong crowd'. As she expressed it: 'I just didn't care anymore'. She then described the school leaving process:

> Just older kids. That's how I met my husband. I made it through ninth (grade). I got a letter that said that I made it through to tenth (grade). But I never reported to tenth. These were about seniors, but they dropped out too. So, they didn't go to school.

I just filled it in with the younger people. We were into partying and drinking . . . all day.

She consequently rejected family and school adults, as well as her school peers; she associated with adolescent pariahs.

Violence

Half of the students did not feel safe at school, and fights and drugs headed the list of what caused this anxiety. They witnessed and experienced violence in their neighborhoods. An African-American male described his community in these terms: 'I always had to fight everybody; watch your back. Just like a basketball game. Somebody would take the ball off you and fight you 'cause you . . . they didn't have anything else better to do than fight'.

This aggressiveness spilled over into the schools. One male informant recalled, when asked if he felt safe in school, 'No. . . . Because of the fights. It's just something at that time you go through'. A female narrator similarly responded to the same question: 'No. . . . 'cause one time this girl was pushed down the steps and this girl started banging her head off the steps and blood was just going everywhere. Kids were free to walk around with knives and weapons'. One male informant recounted outright extortion: 'First semester I was doing real good. Nobody was bothering me. This guy was messing with me and asking for money and I said no'. This student continued by describing how he was further victimized by school administrators, who confused the victim for the perpetrator and punished him for a violation he did not commit: 'One time I found a bullet in my locker. They sent me home for three days'. This persistent harassment by his peers, exacerbated by seemingly arbitrary administrative punishment, led to his decision to leave school: 'I was bullied too much and didn't go to classes'. Another male narrator, when threatened with a knife, defended himself with a weapon; he was expelled.

Many of our informants described humiliating acts and disdain for their peers. An African-American student recalled how students stole her clothes during swimming class.

They took everything I had. I was hurt. I didn't want to go back to school, but I did. I was mad. I don't even want to think about that. I called my mom. They [school authorities] gave me transportation to get home. I had my coat. It was in the winter time. I was cold going home.

She recounted that students sold these stolen items, and those of other students', on school grounds. Another female narrator, who saw herself as a 'nerd', preferred to be with students like herself, 'nice and quiet'. She generally referred to other students in unequivocal terms: a 'bunch of animals'. She remembered peers 'screaming and hollering, also in the classroom. There was never no peace'. This social scene caused her to leave school, since 'there is no discipline'. And when asked if she had it to do all over again, she quickly retorted, 'Probably. The way it is nowadays, you can't even do your work or nothing'. At least two other narrators reported that their schoolbus passes had been stolen, forcing them to walk to their schools. Finally, a female student complained about 'hassles' on the bus and recalled how, while on that bus, 'one boy poured pee on me'.

Such brutal and insensitive acts have become more prevalent in Pittsburgh's schools in recent years. A drive-by shooting, which appeared to be gang-related, occurred at Schenley High School in July 1993 during summer school classes. Several fist fights, according to one student, had transpired several days earlier, foreshadowing this incident. Two days after the shooting, many students, 'deterred' by this violence, remained at home. With two students wounded, they chose safety over schooling ('Schenley Shooting Suspect Arrested', 8 July 1993, Section B, p. 1). In September of that same year, a fight involving twenty students broke out in the main hallway of Allderdice High School, thought by many to be the city's 'top scholastic high school'. This followed a raucous 1992/93 school year there in which three serious attacks had occurred, resulting in one fatality ('Neighborhoods Clash in School', 9 September 1993, Section B, p. 1). One African-American female informant attended the school where that fatality occurred. She noted that other students brought guns with them to school to defend themselves.

The violent acts occurring in Pittsburgh's high schools do not represent isolated incidents, but rather appear to be part of a growing nationwide phenomenon. The headline for *USA Today*'s cover story, on 3 June 1993, screamed: 'Kids, Guns: "It's Shoot or Be Shot"' (Stone, 1993, pp. 1 and 2). This article, which cites statistics and anecdotes from the Department of Justice and the Center to Prevent Handgun Violence, describes how violence plagues schools in small towns as well as large cities. Some 100,000 students carry guns to school every day while another 160,000 remain home because they *fear* school. Drugs and gangs, romantic spats, fights and disagreements, thefts, accidents, and parents modeling violent behavior appear to be related circumstances. The number of deaths for school children has tragically increased from 3088 to 4854 between 1985 and 1990, a 57 per cent jump (*ibid*, pp. 1–2). Most youth violence in American society today is attributed to gangs, but much of this violence, according to our narrators, transcended gang activities.

This issue, in fact, never arose through the majority of our interviews from 1986 through 1991. After that point, however, informants increasingly mentioned gangs and gang influences in their schools and neighborhoods. They all feared gangs. A White male narrator stated that he was 'scared to go to school'. Some of the school leavers we interviewed claimed that gangs did not openly clash in the building during school hours; rather, they confronted each other at school functions or parties. However, an African-American male painted a different picture: 'Too many gang members. You can't walk the halls by yourself . . . You got slit . . . You'd be cut . . . There'd be a couple of gang members down there. They'd want to fight or something'. He thought they were 'stupid: they fight over colors' (of clothing). Nevertheless, because of their presence, school did not represent a good place to learn for him: 'I don't think so. The gang members run that school. There's too many gang members'. Another African-American male echoed his comments, describing the daily routine: 'You got three types of gangs sitting in one (class) room. You know that is going to cause conflict. The teachers are scared of them'. Gangs directly led to his abandonment of school, because he feared for his safety. Gangs, in sum, intensified the violent atmosphere of high schools, and students, fearing for their safety and lives, sought refuge by abandoning them.

A Personal Perspective

An African-American female informant, who labeled herself as a 'good student', was appalled by all of the violence she witnessed and experienced at her former high school.

The kids would like to start fights with you. They would like to test you. They punched this one boy in his stomach and threw him in the garbage can and rolled him down the hallway. I didn't like that. I feel that everyone has a right to come to school and not be afraid. I feel you should be able to come and go secure. I didn't like that.

She simply 'hated' her former high school: 'The kids were just . . . I couldn't believe how bad they were'. She witnessed fights in the lunch-room and bathrooms. This rowdiness overflowed into the classroom.

> You couldn't even learn in the class. With all the boys talking and everything, you couldn't hear nothing the teacher was saying . . . Looking under girls' dresses and taking the mice out of the cages. The teacher can't even talk. She was screaming so loud. They would look at her and keep right on talking. I couldn't learn anything. I'm sitting right in the front row.

While she boasted about maintaining an 'A or B average' in elementary school, at the high school she lamented: 'I messed up'. She claimed few friends because they either rejected her or she could not trust them: 'Some kids refuse to be your friend, they will try to make trouble for you. If you walk down the hallways, they would try to trip you'.

In addition to social isolation, her academic success suffered. She loved many subjects, especially math, but her peers created too many distractions.

> All the things that they were doing really irked you. You can't do your best in class. I can't concentrate or focus on my work. They are busy screaming and nagging at you. I would ignore it, but I couldn't do as well as I could. I need peace and quiet.

She, in sum, felt neither safe nor successful in her former high school. This environment lacked the basic elements of trust: 'You had to watch your back. In these days, they carry weapons — guns . . . They will jump you in a minute. Sometimes they put a towel over your head and beat you up'. This insecure and violent atmosphere produced detach-ment. She felt alone — often preferring solitude. Her academic success declined, and her concerns for her safety intensified. She resolved all of these problems and fears by terminating her schooling. When asked why she left school, she succinctly responded, 'because of the kids'.

Substance Abuse

Substance abuse, according to one African-American female student, represented a serious and pervasive problem at her former high school: 'This is what really messes up high schools, the drugs and booze and cigarettes, all that stuff. 'Cause you figure, the high school kids doing it, the middle school kids are doing the same thing. They can't cope with the dope'. In retrospect, she reflected on the consequences: 'High school dropouts do this, right? Been into drugs, booze, cutting class, getting high, it's great. But you look at it now and it's not'.

Drugs, on the one hand, frightened some informants, as one recounted:

> I had one friend who was messing with this one guy . . . and the guy gave him some cocaine and it was uncut and he snapped. He snapped out. Like the paramedics came and got him out of the sixth period biology class. He was jumping up and down, like he won a million dollars. It was really funny at the time, 'cause we didn't know that he was really flipped out. When they came and took him out in the wheel chair, it was very scary.

On the other hand, drugs appeared to be part of a casual, daily routine for others:

> During lunch, everyone would get high in the back along that stairwell, three flights up the stairwell. I'm now realizing . . . how can they go back to class and learn? The mind is somewhere else. Some went back in class and some just left school and some hung outside of school in the football field.

A White female corroborated this perspective; 'They would sell beer right on the walk. You could go into the ladies room and stuff and smell reefer. I learned that if you put beer in your locker, you could sell it at school. You could get $1.50 to $2.00 for it'.

One dropback reported that peers did not represent the only source of drug use. 'I smoked weed, not now, but I smoked it . . . It started at home . . . Whatever you see your parents do, you want'. She was determined to establish a different model for her own children.

> That's how kids get started drinking and all that. I wouldn't let my kids see me doing that . . . If you see your mother rolling some reefer, and it's laying on the table in this box, wouldn't you try one day to figure out what it was? I tried it . . . I seen them do it. Whatever you see your parents do.

Racism

The topic of racial discrimination remained elusive, if not ambiguous, at times during the interviews. Most informants failed to describe social or systemic racism, seemingly lacking the social consciousness of the problem. However, a few did. An African-American male narrator recognized it and responded to the question about racial discrimination immediately:

> Heck, yeah. Not too much on like the average person's level. But when you get up higher, to those White people who really got the money, it matters to them. I think a lot of them are from the old school. They still have the old school values. They are scared that if Black people take over the country they will try and make White slaves. People are crazy. They are scared of that so they try to keep a Black man down.

This represented a multilayered response. On one level, this student pointed to a relationship between social class and racism, that is, only rich and powerful European Americans fear equality. This implies, on another level, that less affluent European Americans are not racist. Finally, he sees racism both as anachronistic and irrational. As naive as this description seems, it represented one of the few insights we received into broad, deep-seated racism in American society and schools. Only one dropback, an African-American female, stated her view in emphatic and unequivocal terms: 'I believe you have more advantages of being White, 'cause this is a White man's world'.

Otherwise, most informants acknowledged that racism existed in the school setting. Some of them pointed to biased teachers and administrators. An African-American male drew a direct connection between staff racism and his decision to leave school: 'That's one of the main reasons I dropped out of school, because of the racial problem with the teacher'. He never elaborated his comment. However, for him, race mattered in American society, 'because some jobs they judge you by your color. You can't live in certain cities or towns. When you are the only Black or White person in town, you will have problems'. One African-American student waffled on this point, blaming her own race at first: 'A lot of Blacks, they give us a bad outlook some kinda way. When something gets stolen, it's the Blacks. Somebody gets shot or something, it's a Black'. She continued in a conciliatory tone, however:

> We all be treated the same: half the Whites is just as hungry as the Blacks, half the Blacks is just as rich as the Whites. Like there might be a higher percentage of colored students who drop out of high school, but there is also the Whites who drop out too. There might be a higher percentage of White folks

who commit suicide, more than the coloreds, but we all have
our troubles.

A White male student, like other narrators, denied that outright racism
existed, but his comments implied racial tensions: 'There is a lot of
people who are still prejudiced against Blacks. And there's a lot of
Black people who think all White people are prejudiced'.

Other narrators pointed to racism among students, but not in the
usual sense. One African-American female student faced segregation
from her peers: 'As far as females . . . O.K. . . . I had more White friends
than Black. Most of the Black girls, if they see you with White girls,
they wouldn't want to be your friend'. This same narrator maintained
a complex, if not naive, sense of racism, not recognizing it until busing
occurred: 'Black kids had to be shifted and everything. More so the
parents felt that way'. At this point, she stated her position, one she
had earlier alluded to during the interview: 'I always wanted to go to
school with White kids, because Blacks are troublemakers, you know.
They start unnecessary trouble. I felt more comfortable around White
students'. However, she perceived White parents, unlike their children, as hostile to desegregation: 'When they started (busing), it was
more the parents, cause when we was getting shifted from the Hill
District to Mount Washington, our bus was ambushed three times.
The parents in that neighborhood would stop the bus and use clubs
and things and beat the bus. They beat up a couple Black students'.
Her concluding comment proved to be an understatement: 'That was
the only problem we had'.

An African-American male informant recounted similar experiences and feelings, but in more detail. He had attended elementary and
middle schools outside of the city.

> I was in a culture shock. There was a lot of differences between
> coming from the country and being in the city. A lot of igno-
> rant children didn't understand that. I was considered square,
> White, cause I talked fast and I talked proper. Students at that
> time didn't understand that. Because of the students and their
> influence and their negative vibes to me, I started not going to
> class. I was never dumb in class. I just didn't go because of the
> influences of the other students. I didn't like being picked on.

He recalled a specific event in the school's library: 'I was very different.
I was in the library and I was writing a poem. I was singing to myself.
They started teasing me about that'.

In spite of his feelings of rejection, this perceptive student recognized that race mattered in American society.

> Number one — economically, we are not part of the society. We are a subculture. We're labeled, even though there is a melting pot in the U.S. We are labeled as disadvantaged, economically. There is a difference between Blacks and Whites.

He continued by analyzing this generalization within the context of his neighborhood:

> I didn't like Homewood and I probably never will. I won't say the people; I didn't like the way I saw things. It was a culture shock, walking across Squirrel Hill streets or walking across Homewood Avenue . . . Surviving under these situations is not the greatest. I would probably have a grim attitude also. One thing I always came into contact with in Homewood was the drug sellers; people that really didn't care. I saw children running barefooted on the ground. It still hurts me that the parks in the city of Pittsburgh are still closed. Homewood (Park) is still closed for repairs. Yet I go across the street on Forbes Avenue and they are playing tennis there. That disturbs me greatly and it makes me wonder what kind of system I live in. Do I really want to be Black? I am not for Malcolm X, but I do want to do something about this.

Homewood represents a distressed community in Pittsburgh, hosting a predominantly African-American population, while Squirrel Hill is an upper-middle-class neighborhood, claiming an almost exclusively White population. This narrator made one suggestion to the interviewer to improve the schools: 'There was no positive role model for me — a lack of Black teachers'.

Nevertheless, because of the conflict, teasing, and occasional fights he encountered at school, he avoided peers both there and in his neighborhood. 'I just basically stayed to myself. There's a hill behind Crescent Elementary (School). I'd just sit up there. In the summer, I would stay home and sleep. I didn't stay out in the street or on street corners . . . I stayed to myself'. He also expressed regret over not attending a White school: 'The only time I didn't have any problem with people was when I went to predominantly White schools'. He solved his peer dilemma by abandoning school when he was 16 years old.

Fordham and Ogbu (1986) address this 'burden of "acting White"' by explaining the relationship between 'collective identity, cultural frame, and schooling' (p. 180). This represents a long historical process of racial subjugation.

> Subordinate minorities like Black Americans develop a sense of collective identity or sense of peoplehood in opposition to the social identity of White Americans because of the way White Americans treat them in economic, political, social and psychological domains, including White exclusion of these groups from true assimilation. (*ibid*, p. 181)

African-Americans have, in response, created an 'oppositional cultural frame of reference' to 'protect their identity' and maintain 'boundaries between them and White Americans'. Given these cultural contexts, 'behaviors, events, symbols, and meanings' determine affiliation (*ibid*, p. 181). 'Therefore individuals who try to behave like White Americans or try to cross cultural boundaries to "act White" *in forbidden domains* face opposition from their peers and probably from other members of the minority community'. This produces a great deal of emotional stress in individuals, either fearing that they have betrayed 'their group and its cause' or White Americans will exclude them regardless of their behavior (*ibid*, p. 182). The formal schooling process, its curriculum and 'standard academic practices', serves a cultural role in this regard:

> School learning is therefore consciously or unconsciously perceived *as a subtractive process*: a minority person who learns successfully in school or who follows the standard practices of the school is perceived as becoming acculturated into the White American cultural frame of reference at the expense of the minorities' cultural frame of reference and collective welfare. (*ibid*, pp. 182–3)

Subordinate minorities, as a result, resist or oppose the social and academic context of schooling. They either ostracize any individuals they view as traitors, or they themselves avoid academic success (*ibid*, p. 183).

A Personal Perspective

A White male informant gave us another elaborate insight into racism. All of this came as a response to a question about neighborhoods:

'Gangs. It's stupid. It's African-American kids killing off each other. It doesn't make any sense. There's *more* hatred coming out of the White community towards the Black community'. He proceeded to build his analysis, although disjointed at times, and spent a great deal of time at it, without any prompting from the interviewer.

Race, however, did not serve as the root of the problem for this narrator. He focused on the lack of jobs and economically 'depressed' neighborhoods, euphemisms for social class differences, which bred crime. He continued:

> There's a lot of racial problems. I can state this because I'm White. I look from one side to another side. I can understand where its coming from. White people have to understand that you can't point the finger and say that Black person is bad . . . You see White trash. You see White scum that don't want to do anything with their lives either. They go out there and rob houses and steal cars. It isn't about White or Black.

Here he substituted trash for poor when referring to Whites. After dispelling stereotypes, his narrative continued to link poverty with crime. He eventually pointed to economic background as the root of racial tensions. 'It isn't about one race being worse than the other because some White people are just completely trash. Then you got the Black people who look down at them. . . . You got a lot of successful African Americans'.

His diatribe culminated by blaming Whites for perpetuating racism: 'I am angry at my own race because I am constantly hearing the negative things. This is ignorance'. His solutions to this problem appeared to be twofold. First, Whites should not dwell on race: 'Maybe there is a problem we're not looking at'. Second, Whites need to understand other cultures: 'It's better to know about somebody else's culture. It's fascinating'.

Conclusions

Our narrators generally did not paint a glowing picture of their schools, and they felt no loyalty to them. 'School membership' results from 'social bonding', which possesses 'four conditions'. A student is socially bonded to the extent that he or she develops a rapport with adults and peers, accepts school norms, volunteers in school activities, and believes in the legitimacy and efficacy of the institution (Wehlage *et al*, 1989, p. 117). If students fail to develop any sense of loyalty to, or community within, a school setting, then they will abandon it (Roderick, 1993, pp. 90–1).

High schools are not conducive to feelings of membership. Declines in academic achievement usually accompanies the transition from elementary to middle, and from middle to high, school, as we shall see. Couple this with the difficulty of adjusting to different and wider social groups and larger and more intimidating buildings, and students begin to withdraw. 'The failure of a student both academically, and perhaps socially, in these new environments may then initiate or accelerate a process of disengagement from school, and a conflict between the youth and the school as an institution' (*ibid*, pp. 92–3; see also Wehlage *et al*, 1989, pp. 122, 124, 126–7 and 130–1). Violence, drug abuse, and racial tension only exacerbated this situation for our informants. School personnel seldom helped students to overcome these physical and social problems.

Notes

1 These data came from 'Pittsburgh Public Schools, School Profiles: School Year 1992–1993', Division of Student Information Management, January 1994.
2 The information on class size is found in 'Pittsburgh Public Schools Secondary Schools Annual Load Summary Report, 1993/1994 School Year', Division of Student Information Management, May 1994.
3 The data in this paragraph came from 'Pittsburgh Public Schools Student Transfer and Mobility Rate Report, 1992/1993', Division of Student Information Management, December 1993.

Chapter 5

School Personnel

They used to yell all the time, and cuss people out. (Pittsburgh school leaver)

Effective-schools literature stresses six characteristics for a successful learning environment: 'a principal with strong leadership in instruction; teachers with high levels of expectations for student learning; school instructional climate; increased time on instructional tasks; regular and systematic student evaluations; community support and adequate resources' (Spring, 1991, p. 102; Borman and Spring, 1984, p. 202). Students, especially urban minorities, represent the focus in this setting; all personnel should maintain high expectations for their students and coordinate their efforts, and have access to all of the resources they need. However, what we have learned from our interviews and from narrators' descriptions themselves is that, few of these elements existed in Pittsburgh's secondary schools.

Administrators

Principals

The principal in the Pittsburgh schools officially serves as an instructional leader for the entire school building. Our interview instrument did not explicitly inquire about the principal's role, particularly as an instructional leader; students would probably not even understand the concept, nor were they in a position to judge a principal. That would more appropriately be a question to ask teachers. In any case, student responses noted a distinct absence of the principal in anything remotely related to instruction. Students generally viewed principals as 'nice', but beyond that superficial characterization they usually referred to them in shadowy terms. No student, for example, could describe in detail what principals did. Students in many cases did not even know the name of their school principal. When asked to name the principal, most could not. Those who remembered their last building principal were unable to make any evaluative comment. However, students usually associated principals with discipline. It proved difficult, at times, to ascertain from some interview transcripts whether students distinguished between principals and vice-principals, who are responsible for enforcing the schools' disciplinary codes. One male student recalled: 'In the course of my cutting class, I had some run-ins with him'. A female informant saw her highly visible principal 'maybe three times a day' in the hallways: 'He would stop fights. He would stop all kinds of things in the halls that the security guards just by-passed'. The closest response to an evaluative comment was 'strict but nice'. Another student perceived her principals in the following way: 'They seemed to be hidden'. Thus principals, in these students' eyes, failed to play an academic leadership role, reacting to rule infractions or remaining seemingly invisible. How accurate are these perceptions? Students primarily interacted with teachers and other school personnel. Principals, setting the climate of a building, often worked behind the scenes, largely unseen.

Finally, whether a principal was White or African-American, did not seem to affect how students evaluated the principal, but caring and justice did. One African-American female dropback stated that she liked the White principal at her high school because he was highly visible, especially at sporting events. He, in this way, demonstrated that he cared about his students. However, she strongly disliked the African-American principal who replaced him because he only appeared at selected sporting events. More important, he seemed to be 'prejudiced against his own'. She added, 'Instead of taking *all* bad kids, he would take all of the bad *Black* kids. I mean there were bad *White* kids too'.

Vice-Principals and Deans

Other administrators, like vice-principals and deans, fared even worse. Vice-principals typically play a disciplinary role. Although the building principal must concur in any disciplinary action that results in suspension from school, the vice-principal is the administrative authority who makes the preliminary assessment and judgment. The role of a dean in a large middle or high school is virtually the same as that of vice-principal, but is assigned responsibility for a smaller number of students. Many of the informants we interviewed had direct encounters with vice-principals over disciplinary matters and thus received a clear perception of their role. According to one female student, 'he like suspends people . . . They send you home for three days. It stays on your record and all that stuff'. A male narrator echoed her comments: 'They suspended people all the time. That's all they did was suspend people for stupid reasons'. They cited a long list of possible rule infractions, such as not carrying hall passes. As one student humorously recalled: 'He go around and see who is out of class without a pass. Half the times, the boys up there, they had him running after them'. One vice-principal appeared mainly concerned with drugs: 'He was a narcotic man. He tried to keep up with the kids who sold drugs and stuff'. Another male informant, in a more sobering tone, referred to a male dean as 'a threatening type of authority' but remembered a female dean as 'the one I could talk to'. A White female narrator clearly recalled the tasks and procedures the vice-principals at her former high school followed.

> They took care of the kids who didn't like to go to school. If you missed the day, kids would sit in their office and make them do their work there. He used to discipline us and if we cut, they would suspend us and if we fought, we got suspended; normal routine things that vice-principals are supposed to do.

Most of their comments, in the main, indicated that the vice-principals acted fairly.

Nevertheless, some informants were ambivalent about their vice-principals. An African-American male student remembered the arbitrary actions of one of his vice-principals: 'He was mean. . . . If you were walking down the hall the wrong way or clowning around, that would be a write up or detention. If you would say something bad, not a curse word, that was detention, in-school suspension'. He also mentioned another vice-principal, who 'was understanding'. Yet he even qualified this: 'It was just like wild. I had problems. He would say that he would straighten it up. He would never get to it'. For this school leaver, vice-principals either overacted to or benignly neglected students. Finally, in another case, a narrator complained that the dean made him hate school.

Personal Perspective

An African-American female asserted that she was treated unfairly by her vice-principal. First, she described the intense racial antagonism at her former high school, located in a predominantly White neighborhood:

> White and Black students don't get along. We all know that.
> And it just jumped off a big riot. They had chains and pipes
> and stuff like that. We had school trippers; we got bus passes
> and stuff. And so I was on one of the buses. Me and my cousin
> were real close. She was outside in the crowd. They were
> swinging pipes and stuff like that.

Given this menacing situation, she attempted to rescue her cousin during
a violent incident: 'All I was asking the security guard was to let me
out of the bus so that I could snatch my cousin by her ponytail and put
her on the bus with me. He wouldn't'. She faced a profound moral
dilemma: whether to abandon a loved one in a dangerous situation or
physically defy school authority. She chose the latter: 'When you care
about somebody, you don't want to hear what nobody else has got to
say. So, I just moved him out of the way, and got off the bus, and got
my cousin'. That incident occurred on a Friday afternoon. When she
returned to school on Monday morning, she had to meet with the
vice-principal in his office. She described that meeting as confronta-
tional and the vice-principal as irrational: 'I was talking. I was just
telling everything that was going on. He stood up and told me to shut
up. I let him know my mother don't tell me to shut up, and it's not
polite to tell people to shut up, you know'. He ordered her out of his
office, and suspended her for fourteen days, when ordinarily a three-
day suspension was the standard punishment.

Counselors

Although counselors in the Pittsburgh schools must complete graduate
study in counseling psychology, along with a teaching certificate, most
narrators saw counselors as ineffective. When asked what counselors
did at their schools, a few interviewees failed to remember ever meeting with one: 'I can't even answer that question, cause I didn't have
nothing to do with no counselors'. Counselors, seemed to play bureaucratic roles, sorting students. Later, this respondent related: 'They
worked on schedules. Matter of fact, that's the only time I seen my
counselor'. Another student related how counselors in particular and
the institution in general responded to trouble bureaucratically:

> If you kept coming in late, or cutting classes, they would send
> you down to see the counselors, to see if there was anything
> wrong with you. If there was a reason why you were cutting

classes and didn't want to be in school, they would send you to the social worker and determine what would happen and they would send you to the dean and they would take responsibility.

Deans, as mentioned earlier, were usually disciplinarians. Hence, schools appeared to treat these at-risk students as strictly disciplinary problems. One student played the game, however. In his many encounters with counselors, he recollected: 'Those contacts . . . I learned to manipulate. I learned to tell them what they wanted to hear. I don't think they never really understood me . . .'

Narrators, in a few cases, noted counselors who served as supportive advisers, making comments like: 'a good ear', 'always talked to me if I had a problem', and 'helpful'. One female informant reported that she met with her counselors 'about once a month'. A male student likewise recalled positive experiences: 'I can give them credit. They were very understanding, and they helped me. If I had problems with classes, they would switch me around. I had many problems with that. Most of my problems I told my counselor'. A female narrator too observed counselors who, although well-meaning, seemed insensitive, superficial, and ineffective:

They tried to help you, I don't know. They talk to you like you are a dumb person. Get your diploma. Don't worry about kids here. She kinda influenced you in such a good way. That made you have the outlook that I can do it. But when you go out there and try to pull that rope, it seems like that rope is being pulled by a Mack truck on the other side. When you are all off by yourself pulling that little rope, it seems hard, very hard.

These students' perceptions of counselors appear to be universal. Bennett and LeCompte (1990) point out that the 'discrepancy between expectations and actual accomplishments probably is greater for counselors than for any other school professionals' (p. 144). Their responsibilities have evolved to encompass the academic, social, emotional, and psychological needs of students. Counselors spend 75 per cent of their time on paperwork, such as maintaining student records and preparing schedules (*ibid*, p. 145; Powell, Farrar and Cohen, 1985, p. 49). To compound their already complex job, counselors nationwide carry a difficult work load, with 350 to 420 students. Pittsburgh's counselors typically oversee 300–500 students. Students therefore seldom can arrange an appointment with their overwhelmed counselor, and the average office visit lasts less than ten minutes.

> The result is that only those students with problems — pushy parents, personal persistence, serious discrepancies in their records, or visibly flamboyant misbehavior will receive attention. Finally, students do not get placed in programs appropriate to their capabilities and minority children fail to get encouragement and guidance they need to stay in school and aspire realistically to college. (Bennett and LeCompte, 1990, p. 145)

Lee and Ekstrom (1987, p. 288) explicate this process in detail. Relying on data from the extensive *High School and Beyond* study, they focus on 'academic counseling', that is, scheduling and program planning as well as career and post-secondary education advice. Guidance counselors have long played fundamental roles in students' academic growth and development, serving as 'critical "gatekeepers" in students' progress through the educational pipeline'. Three universal and serious problems inhibit students 'access to high school guidance counseling and the type of counseling students get'. First, student-to-counselor ratios remain too high, as we have seen. Second, counselors appear to be preoccupied, if not overburdened, with bureaucratic 'record keeping' tasks. Third, 'counselor training derives from clinical psychology and has as its goal "developing the whole child" rather than mediating "how students use a choice-based curriculum"' (*ibid*, p. 289). Thus counselors cannot help students with current program planning and with post-secondary plans.

Few students receive advice from guidance counselors about their academic careers; 'slightly more than half (50.4 per cent) of high school sophomores report no contact with a guidance counselor when making these important decisions' (*ibid*, p. 296). When Lee and Ekstrom (*ibid*, p. 298) broke down the data by socioeconomic background, race and ethnicity, the picture became clearer: poor minority students seldom have counselors. Counseling here serves as a 'causal factor': 'Students who have access to counseling are more likely to be placed in the academic track (Signifying, of course, that those with no such access are more likely to end in non-academic tracks.)'. Lee and Ekstrom (*ibid*, p. 300) stress that counseling functions as a 'social stratifier in our nation's public high schools'.

A similar pattern exists for advice on post-high school plans: 44 per cent of students report that no counselors advised them about their future educational or career goals. Those students lumped into the 'general curriculum track' rarely have access to counseling services (*ibid*, pp. 304–5):

> This finding is particularly troubling, since the general curriculum appears to be less focused on post-high school outcomes than either the academic curriculum track (which is designed to prepare students for employment or additional vocational training).

The irony here is that the students most in need of academic counseling fail to receive it.

An African-American male recognized differential treatment given to students according to academic ability and career goals.

> I was under the impression that maybe the counselor was too busy. He always seemed like he was in a rush to do something. You got the feeling that if you wasn't a scholar, they wouldn't take that much time out for you. If you were going to go to college, they would take time for you.

The 'scholars' appeared to receive preferential treatment by everyone: 'If you were a scholar, you were somebody. They went on a lot of field trips, the scholars. They used to go on the bus everyday'. The 'scholars program' stresses an intensive approach to academic work.

In sum, counseling appears to be available, but not on an equitable basis: 'It seems that disadvantaged students are either systematically or unwittingly being excluded from an important mechanism in the nation's public high schools, a mechanism they may need even more than most students' (*ibid*, pp. 306–7 and 308).

Teachers

The subject of teachers and teaching generated the strongest responses. Our interviews demonstrated how little we had appreciated the critical role that teachers play in the lives and decision making of these school leavers. Informants defined 'good' teachers by using the following descriptors and phrases: 'helped', 'work with students', 'allowed conversation', 'explained', acted 'like a big brother', and 'keeps on you' (that is, makes students complete their assignments). In most cases, they only identified a few good teachers. The preponderance of comments which indicated that such teachers 'helped' or 'explained' or 'made things clear' or 'worked with students' indicate a high level of involvement that facilitated such positive responses. Similarly, the comment that a good teacher 'keeps on you', while suggestive of strictness, shows that such attention was apparently appreciated.

Most students had only negative comments to make: a poor teacher did not help, did not explain, ignored students even when their hands were raised, had favorites, did not care, was 'mean', or 'picked on' students, 'tried to rule you', treated students 'unequally', did not understand, and was just 'talk, talk, talk', 'talked all period'. A common perception was of distance between teachers and students. We have no grounds for evaluating the attitudes that students brought to their relationships with teachers. Were they initially hostile to any classroom environment, to all teachers, to their peers, or all three? Such possibilities cannot be excluded. A male dropback even blamed students for teachers' poor attitudes: 'The kids did whatever they wanted. They would talk and mess around in class. They walked out of class. They did whatever they wanted'. On the other hand, he blamed certain teachers for classroom disruptions:

> Well, some of the teachers had a way. You are not going to do that in my class, and (the students) gave the respect to some of the teachers. Some (teachers), they didn't give respect to students; the kids would run over the person. . . . They would disrupt the class.

Informants clearly felt antagonism with most of their teachers. Comments such as a teacher 'not caring' or being 'mean' or treating students 'unequally' or picking on students indicate a profound lack of rapport between teachers and students. How that gulf can be closed is a relevant question. Further, the comments, such as a teacher being only 'talk, talk, talk' or that a teacher 'talked all period', indicate a style of pedagogy where classroom activity revolves around what the teacher does with insufficient attention to what the learner does.

Teachers, therefore, represented the best and the worst of schooling for our informants; they helped to shape their most and least favorite subjects. In other words, students identified their subject matter preferences based on their perceptions of teachers. As one female narrator summarized it, some teachers are 'real nice', but 'some teachers are ignorant'.

Students sorted teachers according to surprisingly simple criteria; race appeared to mean little, but caring and pedagogical skill did. Caring, a subjective or affective element, implied selfless, egalitarian and consistent treatment of all students. One African-American female informant easily and clearly described a caring teacher: 'Someone who takes time and who sits down and helps you when you need their help and who will give you their attention that you need . . . Two that I can really remember; that really stuck out at the time'. On the other hand, she described poor instructors as 'teachers that really did not care. Who have the time to spare but is not willing to give the time to you. I had a lot of them'. Through many of our interviews, caring for students appeared to be consistently centered on two basic elements, sharing time and exercising patience. Our narrators appreciated those seemingly selfless classroom instructors, who, devoted to a nurturing process, freely gave their time reviewing or reteaching the information. These were too few. Other studies confirm this finding. A mere 6 per cent of potential dropouts view their teachers as 'friends' (Beck and Muia, 1980, p. 69).

Many of our informants also recalled favoritism. One narrator described a humiliating experience for many students:

> Everytime after a test . . . she would pass out the tests and read out your scores. The perfect students would be read off first, and there would be like three As in the class, everybody else are Ds and failed. I guess the teacher did that to make the people embarrassed to make them try harder.

Students either resented it or appeared resigned: 'All teachers have their favorites. I think so. But there is one teacher . . . that don't though. She treats everybody the same. That is what she should do. Some teachers, if they don't like the students they show it'. One female student observed how teachers chose favorites:

> Most of the kids that were real smart in class. The other ones, they just ignored altogether. They was always nicer to those (smart) students and always mean to the others. They would like, if a kid missed a day of notes, he would give it to him and help him out, but he wouldn't the other students.

Another student stoically responded: 'It didn't bother me'. Kelly's (1993, p. 105) interviews of California continuation school students find similar treatment based on academic achievement and ability.

Pedagogical practice represented a more observable and less subjective measure of teacher effectiveness, and students maintained a critical and perceptive eye concerning the classroom routine. Too often we heard: 'Some (teachers) don't care if students learn'. Teachers, concerned only with control, combined discipline with pedagogy, as in the case of one 16-year-old school leaver: 'You would be in class and they would give you an assignment or something. If the class is disruptive, they would give you a test on it, and expect you to pass it in a few minutes!' The failing grade assuredly received through this process served at best as a lesson about behavior, and operated at worst as punishment, giving the students what they deserved because they were off-task. However, not all informants criticized this dreary classroom environment. One student saw social studies as 'fun' because, as he stated it, 'all we did was dittos and look the answers up in the book. And we got good grades'. This narrator appreciated his teacher's lack of intellectual challenge, as long as he received a 'just' reward. Whatever the case, students grew detached and felt unchallenged.

These do not represent new experiences. Cuban (1984, p. 2) argues that secondary teachers have not profoundly altered their pedagogy since the late nineteenth century, despite intense reform efforts to move classroom practices toward instruction that was more student-centered. In learner-centered classrooms, students have opportunities to talk as much as, if not more than, the teachers. Students often work individually or in small groups, seldom as a whole class. Students also assume a great deal of responsibility selecting and organizing the content of what they learn, as well as creating and enforcing classroom rules and procedures. Learner-centered classrooms too contain a variety of instructional materials and resources, and students have access to these for at least half of the time. Finally, classroom desks and other furniture appear to be arranged to facilitate individual or small group work (*ibid*, pp. 3–4).

This setting sharply contrasts with teacher-centered classrooms, which our narrators consistently described. Teacher talk dominated the time, usually during whole-class sessions, with the desks arranged in the traditional rows facing the teacher, the teacher's desk, and the chalkboard. Teachers, in other words, controlled every aspect of the classroom, including how the time is utilized and the furniture is arranged (*ibid*, p. 3). Unlike, student-centered classrooms which function as democratic settings, with shared decision-making and concerned with learning, teacher-centered classrooms were authoritarian environments, emphasizing unilateral pronouncements and focusing solely on teaching as telling. Since students play important roles in the former, they naturally place a higher value on learning and appear to be more motivated and intellectually engaged.

Change has been slow in coming for several reasons. First, schools have long stressed social control and have functioned as mechanisms for sorting students (Spring, 1976). As Cuban (1984) explains it, recapping this twentieth-century experience,

the ways schools are organized, the curriculum, and teaching practices mirror the norms of the socioeconomic system. Those instructional policies that seek obedience, uniformity, and productivity through, for example, tests, grades, homework, and paying attention to the teacher prepare children for effective participation in a bureaucratic and corporate culture. Consistent with this argument is that certain teaching practices become functional to achieve these ends: teaching the whole class encourages children to vie for the teacher's attention and encourages competitiveness; teacher questions reward those students who respond with the correct answer; rows of chairs facing a teacher's desk produce a uniform appearance, reinforcing the teacher's authority to control the behavior of the class . . . Teacher-centered instruction endured because it produces student behaviors expected by the larger society. (p. 9)

Second, school administrators have emphasized efficiency and management, and teachers over time have adapted approaches that achieved these bureaucratic imperatives (Callahan, 1961). 'Lecturing, numerous teacher questions, and seatwork are simple ways of conveying knowledge and managing a group efficiently'. Third, the teaching culture has maintained a conservative, if not rigid, stance, shunning innovation and preserving tradition. 'People who become teachers . . . themselves watched teachers for almost two decades before entering their own classrooms. They tend to use those practices they observed in teachers who taught them' (Cuban, 1984, p. 10). This vicious circle has ensured a stagnant but constant pedagogy. Many students refuse to accept it, however. Fourth, this pedagogical process has functioned as a result of the tacit 'treaties' commonly negotiated in high schools, where teachers have demanded little and students have given even less (Powell, Farrar and Cohen, 1985, pp. 68–70). This subtle, but profound, disengagement from learning exacerbates student alienation.

The theoretical grounding for such teacher-centered pedagogy seems to be a crude form of classical learning theory which posits that 'experience is something that happens to a learner'. More recent learning theorists, in contrast, argue that 'experience is something that a learner engages in, it is something that *transpires as a result of the interaction* between a learner and the surroundings' (Phillips and Soltis, 1985,

p. 17, emphasis added). What would have engaged the school leavers interviewed? Some narrators provided us with valuable insights. When we pressed students to describe their best instructors, one female recalled how her favorite teacher taught her about Shakespeare and his works:

> I liked *Hamlet*. All the ones like that. I can't remember all of them. We had so many. We had a good English teacher . . . If kids would have had her, they probably would have liked it. It all depends on who you have. Some people don't give a darn. They say 'read this'. She was like exciting. She was really into it. If you didn't understand it she would help . . . She would act out a part. Everybody liked her.

Another female informant defined 'good' teachers as simply helpful and patient: 'They helped you. If you had a problem, they would help you. They would show you as many times as it took. Like math, the teacher would help me with math. He showed me as many times as it took and he wouldn't get mad like most teachers'. Therefore, according to our narrators, simple ingredients, like obvious caring or creative pedagogy, defined 'good' teachers.

Gender differences may encompass both of these variables. According to Brophy (1985, pp. 122 and 136–7), the classroom routine often reflects the effects of gender: 'Data from several studies suggest that male teachers tend to be more direct and subject-centered and female teachers to be more indirect and student-centered'. Male instructors prefer to lecture while females rely on questioning; males emphasize wrong answers while females frequently praise student efforts; and males respond to student failure while females appear to be warmer and stress success. As a result, Brophy (1985) asserts that

> students had more public response opportunities, initiated more private contacts with the teachers in classes taught by females. They were also more likely to guess when unsure of their answer in female teachers' classes but more likely to remain silent in male teachers' classes (suggesting that the students felt safer in guessing in the female teachers' classes). (p. 122)

Female instructors, as developmentalists, are more apt to emphasize learning while males, more often traditionalists, are likely to stress teaching. Kelly (1993, p. 118) reinforces Brophy's conclusions, because she observed in her California interviews that students in traditional, teacher-centered classrooms abandoned learning even though they continued to attend school.

We did not foresee this pattern of differences between male and female teachers, but it emerged, nevertheless. We found isolated and inferred gender patterns rather than widespread and blatant ones. When we probed them, our informants voiced complaints about their high school teachers, but never criticized their middle or elementary school instructors. Male teachers, of course, dominate at the secondary level. We did not quantify narrator's references to faculty according to gender, yet trends did appear. These students consistently made negative comments about 'his class' or 'he', that is, 'poor' teachers generally, but not always, seemed to be males. When our interviewees mentioned 'good' instructors, the pronoun 'she' often appeared.

Students were thus more positive toward their elementary teachers, who were usually female, than their secondary instructors. A female student expressed deep affection for her elementary teachers: 'I loved them all!' She organized a surprise baby shower for her sixth-grade instructor: 'Me and my friends got together. She was pregnant. Her birthday was 6 June. I remember; we all got together, pitched in, and gave her a party . . . I'm the one who got the party started'. Neither this nor any other narrator expressed such ebullience for their secondary teachers. A male informant identified his third- and sixth-grade female instructors by name when asked about 'good' teachers. 'They used to treat me good. They used to be nice to me. They would help me in my work and stuff . . . Drove me home from school; I used to stay after school and wash the boards'. Another male narrator described how one of his elementary instructors sacrificed her personal time for him. 'On off days (she) would come to my house, before my mother went blind, and me and her would go on trips, maybe to the mall. She would later take me home and give me a special spelling test or reading test'. Finally, a fourth student summarized it in this way: (Elementary teachers) 'just cared more and explained. When you get into high school, it is completely different'. This difference appeared to center on the teacher's role, that is, elementary teachers comfortably moved beyond the traditional instructor's responsibilities. They clearly cared about the students and expanded their instructional role to compensate for their needs. These students recognized this nurturing process and appreciated their adult contacts.

Thus, transitions from elementary to middle and high school prove to be more than 'ecological stress', that is, not just adjustment to new social and larger physical environments, but include adaptation to educational practices and personal relationships. One informant described her impressions of the differences between elementary and secondary levels: 'Middle school and high school is completely different. Teachers teach you. It's the teacher'. Roderick (1993) found in her Fall River, Massachusetts, study that

> junior high school teachers were more likely to rely on educational techniques — including public evaluation of students' work, the use of whole class rather than small group instruction, and more rigorous grading procedures — which highlighted and encouraged comparisons of students' abilities, rather than emphasizing individual task mastery and progress. Teachers in junior high school math classrooms placed greater emphasis on discipline and less emphasis on student participation. In addition, students perceived junior high school teachers as less supportive than their elementary school instructors. (p. 134)

These teacher-centered practices intensified at the high school level. The results proved tragic.

> As at-risk students move into middle school and high school, their interactions with school personnel become more anonymous and less supportive, their in-class experiences become less engaging and rewarding, and they receive direct messages in terms of track placement regarding their relative position in their school.

All of this results in student 'isolation from that school community' (*ibid*, p. 135; see also Wehlage, 1989, pp. 10–11). The next step seems to be obvious.

A Personal Perspective

Some narrators reconstructed classroom routines in detail. One female student provided a thoughtful, balanced, and sensitive view, first briefly describing two 'good' instructors, then quickly moving onto her experiences with a 'poor' teacher.

> I liked my gym teacher and my music teacher. Those classes were interesting. The teacher took time with you. She didn't rush you into things you didn't want to do. (Other teachers) just throw a paper in your face and say do it. Whereas some

students may not be on the same level as other students. And they just . . . sometimes they just don't take the time out. Like everybody is on the same level.

She proceeded to recreate this latter process through a description of her US history course: 'I used to like history. In about the ninth grade, I used to like history, but when I got to the tenth and eleventh grade, it got kinda hard. Really hard. It didn't seem like I had the interest in it like I did when I was in the ninth grade or anything'. Her observations pointed directly to the teacher, not the subject matter: 'You know when he talked to us it was really fast and it just seemed like the class . . . I don't know . . . everything was on the same speed, but I wasn't. What he was talking to them, I hardly understood. It seemed he was talking Chinese or something'. This, coupled with the size of the class, intimidated her:

And then the class was so big. We had about twenty-five or twenty-eight people. I really felt like they really think I was dumb. But I'm not. I never really asked him things. Cause the class seemed too big, I would have never been like the only one who raised my hand . . . What I'm saying is that he is only one person. In a big class, like you can't just sit there and tutor one child.

The informant, at this point, ignored the interviewer, who had commented that the class size she described was average, perhaps modest in some cases, and persisted with her recollection, focusing on the instructor and the total absence of class discussions: 'What he said, was said. Then the kids would get to their papers and do their book work. I probably would catch up with the book work. And when it came to him telling us, like a filmstrip we just seen, maybe giving us some background on that, it was crazy the way he talked, real fast'. She consciously avoided going to the teacher for help: 'Sometimes my older sister would help me, cause she would take the same class but different periods. She made me have a little clearer look at it, but by that time I was lost'.

When the interviewer asked whether teachers seemed to care, this student was unequivocal: 'They couldn't care less if you studied or not. They were just getting paid. That's what they were saying to us. Yeah, a couple of teachers told us: "I'm getting paid, so if you want to sit there and joke around, I'm still gonna get my check." That's what they said'. This narrator perceptively recognized and grappled with a contradiction here, first noting the idealism which had originally spurred most teachers to enter the classroom: 'Some of them had a nice thought about kids. For them to be a teacher, they have to have some kind of interest in the child, right? They want to give something to the child that they never had. Some are like that and some of them say that they are there for the pay'. She then attempted to explain such teacher burn out — albeit in vague terms: 'I have to admit that they go through a lot of hell'. But she quickly returned to a critical perspective, offering insight into the harmful consequences of irresponsible teacher comments: 'If teachers say that to a student . . . you say *wow*. Why he say that? That makes a student want to leave'.

Student Actions

Students responded to uncaring teachers and poor teaching in a number of ways. Some accepted it, and continued to attend classes. Others simply cut classes. Cutting was a more subtle form of school leaving, that is, these students, while still formally enrolled in school, began the disengagement process. However, one male student, who resented 'dull . . . and very impersonal' instructors, circumvented the entire system by creating his own schedule which was based strictly on the teachers, who 'were real helpful. The funny thing about either of these teachers, I wasn't scheduled for their classes. I just got a chance to just come on in. I went to the classes that I thought were informative. . . . They were different than most of my other classes'. Therefore, students, through their own informal network, knew which teachers were 'good' and which were 'bad', even before attending their classes, and some adapted accordingly.

Attitudes Toward Other School Staff

We asked all informants about social workers, yet we received few responses. Social workers thus appear to maintain a sketchy role in school leavers' lives and decisions. The only descriptions we received came from female narrators, who painted a positive picture of social workers' jobs and their assistance to students. One saw social workers as understanding and helpful. School social workers 'talked about your problems at home, cause some kids drop out of school because of problems at home, you know. They helped you through them. They talked about what was your home life like. Is that interfering with your school? If it is, how can we help you in some way?' Another informant appeared to be very grateful for her social worker's advice and aid. Her high school's social worker urged her to go to Letsche Education Center, the district's alternative school, to protect her unborn child: 'I didn't want to stay there because of the fights. You lose your baby over standing around trying to get to your class. You get knocked down or something. So, (the social worker) preferred me to go to Letsche'.

Pittsburgh's high schools, like many other large urban districts, maintain security personnel to keep order. If administrators and counselors achieved limited visibility, uniformed security forces maintained a high profile. Informants' comments about their schools' security forces covered the spectrum.

Most judged the security forces to be both 'fair' and 'necessary'. A particularly graphic response indicated that the school was 'a jungle', and therefore security personnel were necessary. Others simply commented that they were 'real nice'. One female saw them as indispensable: 'If you don't have those, I think the school would be terrible, really terrible. (The schools) are already terrible, terrible; they would be terrible. They need more security guards'. She seemed to associate schooling with prison. Another female narrator saw them as protecting both students and teachers: 'I know those teachers can't be handling them, because a couple times a couple teachers got punched in the face breaking up a fight. That's not their job'.

Other informants' comments, however, led to serious questions about the integrity and effectiveness of some security personnel. For example, one student apparently caught in some infraction reported that a security officer said: 'I won't tell if you give me a reefer'. Was this comment manufactured? There is little reason for the student to lie. Another narrator complained: 'Ah man, they was a real pain. They would stay on you all the time. If they see you in the hall, they would want to know everything. What class you going in, or where are you supposed to be at'. He resolved this conflict. 'You gotta leave the school then', he concluded, and laughed.

Other students were ambivalent. One stated that security personnel were 'cool,' yet went on to say that one security person was a 'smart mouth' who threatened, 'I'll kick your butt'. One male narrator appeared to be extremely ambivalent when he described his school's security force. Although the security guards prohibited wearing of shorts and use of tape players equipped with earphones, popular student accoutrements, guards proved useful when 'other people come up to the school and try to start trouble'. While security personnel did not keep students in school, they did seem to do what they were there to do, namely, maintain order.

The subject of the office staff elicited few responses from our narrators. Their recollections were usually positive, describing them as 'nice'. One male student criticized them yet also felt sorry for them: 'They were very unorganized. But I really didn't blame them because you could just see that they were really bogged down. There was a lot of paperwork and like they needed more people'.

Conclusions

School administrators played a role of less obvious importance to the school leavers we interviewed. They did not understand the principal's role. They knew the vice-principal's role only because they directly observed it or experienced it. Whether or not such personnel play a supportive role can be questioned. We talked with students who did not have a positive school experience — they had abandoned school. These school leavers had received no incentive from administrative personnel to remain in school.

After over a decade of reports that highlight the putative failure of public school teachers, it would be easy to reiterate that blanket indictment. But we doubt that teachers represent the single factor in school leaving. Still, if teachers overlook their crucial role in student development, school leaving will surely continue at current levels. Pedagogical methods, especially at the secondary level, need to be reexamined.

Students must be given more support and must become more involved in the learning environment. Otherwise, today's educational problems will become society's tragic dilemmas tomorrow. What stuck us most vividly through the interview process was that not one informant described an adult advocate, that is, an administrator or a teacher, who defended the student's interests and needs in school (Wehlage *et al*, 1989, p. 23). More than alienation from academic subject matter and other educational and social activities in their schools, students felt estranged from the adults who were supposed to help them.

Chapter 6

School Knowledge

You are not going to always find a job without an education.
(Pittsburgh school leaver)

Academic disengagement stems from three experiences (Wehlage, 1989, pp. 180–5). First, the existing curriculum structure in American high schools emphasizes broad coverage, resulting in superficial learning at best. Some students, as a result, do not master skills or gain a deep appreciation of knowledge. Second, school subjects seldom appear relevant to some students. They rarely see any connection to their social realities, and consequently dismiss what they are being taught or exposed to. Third, 'disengagement from academics is "cool" in the peer culture'. Social acceptance, in adolescent culture, often supersedes any hope of academic success. Students thus fail to experience 'intrinsic' and 'extrinsic' rewards regarding what they learn in school (*ibid*, p. 187).

Our goal in this chapter is to reconstruct what our informants learned in school. Powell, Farrar and Cohen (1985, pp. 2 and 13–21) divide this knowledge into four curricular areas in the typical high school. First, the 'horizontal curriculum', characterized by extensive variety, offers different subjects like foreign languages, sciences, math, and English. Second, the 'vertical curriculum' refers to the same subjects, but at 'various levels or degrees of difficulty'. As Powell, Farrar and Cohen (1985) summarize it: 'Some schools speak of honors, college preparatory, general, and basic levels. Others use words such as accelerated and regular, and still others employ letter or Arabic or Roman numerals' (p. 22). Third, the extracurriculum, or co-curriculum encompasses 'sports and other non-academic or avocational activities'. Educators have always considered it 'integral' to the high school experience, supplementing the formal curriculum. This makes it no less educational in their eyes, but it serves, at the very least, to foster self-esteem and operates, at best, as a 'learning experience', defined in its broadest terms, because it is considered as 'life-learning' not as 'book-learning' (*ibid*, pp. 2 and 30). Following this logic, students who reject dreary classroom learning might divert their learning opportunities to these informal activities and, more important, remain in school. Fourth, the 'services curriculum' addresses the emotional, health, social, and physical needs, problems, difficulties, or crises of students, and is co-ordinated by guidance counselors, social workers, security forces as well as school nurses. High schools offer such a comprehensive program in order to, as educators have long reasoned, attract, motivate, and hold diverse groups of students. Progressive school reformers, during the early decades of the twentieth century, attributed school leaving solely to a lack of student interest, and pushed for a more varied program as a solution. Secondary education has tried in this way to meet the 'individual' needs of students (*ibid*, pp. 34–5 and 40).

The first two curricular areas comprise the formal, or academic, curriculum. Our narrators seldom mentioned any affection or predilection for a specific subject area. Nor did they express enthusiasm or positive feelings for their extracurricular activities, refuting traditional rationales for it. Since we addressed the services curriculum in chapter 5, and found therapeutic care to be wanting, we omit it from this chapter. However, we offer two other kinds of learning experience that emerged from the interviews. Our students often talked at length about their work backgrounds and drew conclusions about the relationship between work and schooling. Finally, a very few narrators opened up and described a 'social curriculum', where they learned values, symbols, and knowledge necessary for their sheer survival from their peers, but not for academic success.

The Formal Curriculum

School leavers' academic backgrounds and experiences are varied and complex. As alluded to in chapter 1, a poor academic record coupled with repeating a grade virtually assures that a student will leave school. Citing a Maryland report, Beck and Muia (1980) write that 'half of the school dropouts have been held back at least once' (p. 68). They further assert that, 'in view of similar reports, this estimate is a cautious one'. The typical school leaver's reading ability usually falls two years below grade level. On the surface, this appears to be a logical statement supported by corroborating evidence.

Roderick (1993) contends that more subtle and thicker patterns exist. Her study of dropout students in Fall River, Massachusetts, which traced students from the fourth grade, found that

average grades and attendance of the majority of dropouts . . .
were not substantially different in the fourth grade than those
of students who went on to graduate in the bottom third of
their class. Trends in the average grades and attendance of late
grade dropouts and graduates, moreover, followed quite similar patterns as they moved through middle school and into
high school. (pp. xviii–xix)

However, a common experience among students who left school was
that they found the transition from middle to high school to be a social
and academic disaster. Their grades declined sharply after the transition, and never recovered. Roderick points to a gradual divergence
between graduates and school leavers at the high school level; the latter
group reflected academic disengagement through increasing absenteeism. Those who left school typically had repeated a grade. 'In this
study', Roderick (1993) reported, 'students who were retained in grade
were more than three times more likely to drop out of school than
students who were never retained' (*ibid*, pp, xix–xx). This created low
self-esteem, embarrassment at being overage for a particular grade, and
isolation from peers. Therefore, traumatic transitions and failing to
keep up with one's age group, according to Roderick (1993), marked
the typical school leaver's path. More importantly, this invalidates the
stereotype of the school leaver, commonly categorized, and sometimes
dismissed, as of 'low ability'. 'A case in point is Chicago where nearly
a third of all students who dropped out of the Chicago public schools
in 1984 were reading at or above grade level' (Wehlage *et al*, 1989,
p. 48).

A review of ten of our students' school records both supported
and refuted this account of the disengagement process. This analysis
covered three components: standardized tests, classroom grades, and
attendance patterns. We randomly selected five female and five male
records and found sharp differences between gender groups. All five
of the females who left school read at or above grade level, according
to the California Achievement Test (CAT) annually conducted by the
Pittsburgh schools. Course grades indicated no consistent failure trends;
rather, they portrayed moderate success at best and marginality at worst.
Their attendance patterns appeared to be relatively regular, belying an
initial distaste for schooling. In sum, these students appeared to be competent, and in a few cases highly capable (Voss, Wendling and Elliott,
1966). However, their classroom performance and attendance deterio-
rated just before they abandoned school.

More complex experiences emerged from an analysis of male school records. Although differences existed among all five, the data show two important commonalities: problems in reading and language, and a high absentee rate. First, all of these students experienced deficiencies in reading comprehension and language expression. The achievement test scores for four students correlated with their assigned grades. However, one student's reading comprehension scores steadily declined from third through fifth grades, yet his school grades indicated a passing grade of a 'C'. Second, all five male school leavers experienced attendance problems, but they maintained distinct patterns. In the most dramatic case, one student transferred out of the district so early during first grade that no classroom evaluations were recorded. He later returned to the district, entering the fourth grade, and compiled a record of high absenteeism and low achievement. His fifth-grade experience maintained that same trend. He repeated fifth-grade twice, then was promoted to middle school. In another case, a student transferred into the seventh grade from another state. During the next two years, he attended two different middle schools because of residential moves. The school district's policy allows students to remain at the original school, but this student, for whatever reason, switched schools. His poor academic performance caused him to repeat eighth grade, yet with increasing absences. In his first year of high school, he failed all of his academic subjects, and abandoned schooling.

Unlike the five female students, the seeds of school leaving for all five male students could have been identified during their elementary years. If we assume that reading and writing abilities represent the primary products of the earliest grades, and that later learning depends on these skills, then each of these male students exhibited clear early indicators of later academic problems. The records indicate no intervention occurred in any of these cases. In light of recommendations by urban school superintendents, early intervention represents an unavoidable necessity (OERI, 1987, p. 6 and 16ff.). The CAT, as well as teacher-made tests, could have been used as diagnosis instruments. High absentee rates not only compound these academic problems, but signal gradual disengagement.

However, our interviews refuted traditional perceptions and uncovered a different phenomenon, stressing the affective rather than the cognitive realm. Unlike the stereotype — and what most struck us sixty-five of these young people expressed, in their own ways and words, a consistent interest, even a love of reading. This observation was based solely on their, formal versus informal reading activities. For example, an African-American female informant, early in the interview, claimed that she never seemed to be a good reader in school;

however, later during the interview, in response to a question about casual reading, she claimed to read newspapers and 'spooky' books. Others read local newspapers, like the recently defunct *Pittsburgh Press*, 'to see what is going on', as one student expressed it, or the renowned African-American *Pittsburgh Courier*, 'cause it was about Blacks', as another narrator noted. They 'browsed' through the usual adolescent magazines, like *Sports Illustrated* and *Seventeen*, or adult ones, such as *Ebony* and *Jet*, or atypical ones, like *Popular Mechanics*. More important, many consumed books — albeit largely escapist ones, like Stephen King's novels, as well as solid literature, from authors like Judy Blume and Edgar Allan Poe. They read for fun. A White female belonged to a Book-of-the-Month club. Another informant recounted her reading experiences as such: 'My mother used to get books. One of the first books was *Catcher in the Rye*. My mother introduced me to *Mommy Dearest*, so I read that'. An African-American student recalled a similar pattern: 'I read books . . . (about) Mandela, Martin Luther King, Jr. . . . biographies . . . My mother buys them at the store for herself, and I read them'. Still another read *Malcolm X*. Finally, one African-American female dropback boasted: 'I like books. I love novels . . . detective stories, romance, mysteries, things like that. I like Jackie Collins. I like different kinds of books. I even read educational books, too'. An African-American male informant loved reading so much that he consumed his mother's nursing aid books, his sister's 'college books', as well as his Bible. At least two of our narrators read to their own children. One, who liked to read Shakespeare, read Langston Hughes to her young child.

Few students expressed such enthusiasm for their school textbooks, though, regardless of the subject matter. Teachers and other educators they encountered too often overlooked students' outside entertainment and intellectual interests. These school leavers never recalled encountering an English course that assigned popular novels, or a history class that relied on relevant biographies.

School texts have long been criticized as sterile, dull, and biased (Anyon, 1979 and 1983). The Commission on Reading (1984, p. 65) analyzed the mechanics of reading in American schools, focusing on basal readers. First, textbooks appear to be replete with 'bad writing'. Regardless of the discipline,

> scholars who have examined subject matter textbooks often have failed to discover a logical structure. Sections of many textbooks consist of little more than lists of facts loosely related to a theme. Abrupt, unmotivated transitions are frequent.

Textbooks are as likely to emphasize a trivial detail or a colorful anecdote as a fundamental principle. (*ibid*, p. 69)

Second, supplementary materials rarely prove to be more stimulating, adversely affecting pedagogy. Students spend an inordinate amount of time on ancillary resources like 'workbooks and skill sheets'. Elementary children alone complete 1000 worksheets a year (*ibid*, p. 74). These do not require writing or high levels of thinking, and generally appear to be unstimulating. Close adherence to a textbook approach constrains teachers' responses to 'learner variation' and inhibits the use of creative approaches (Rozycki, 1992, p. 161). Students, as a result of years of lack of motivation, become bored with their formal reading materials.

Michael Apple (1986) adds a political dimension, that is, texts standardize knowledge which is a part of a 'larger system of control' (p. 82). 'It is estimated', he continues (p. 85), 'that 75 per cent of the time elementary and secondary students are in classrooms and 90 per cent of their time on homework spent with text materials'. These required readings define 'what is elite and legitimate culture' (p. 81). For Apple,

the choice of particular content and ways of approaching it in schools is related both to existing relations of domination and to struggles to alter these relations. Not to recognize this is to ignore a wealth of evidence in the United States, England, Australia, France, Sweden, Germany, and elsewhere that linked school knowledge — both commodified and lived — to class, gender, and race dynamics outside as well as inside an institution of education. (*ibid*, pp. 84–5)

Sterilized textbook knowledge, therefore, anesthetizes students concerning inequality and vital social issues (Claybaugh, 1992, p. 160). Given this social and political context, it is no surprise that many of our narrators did indeed read, and preferred biographies about minority struggles and heroes, newspapers, and literature that focused on minorities and minority perspectives. Lack of reading skills did not appear to be as serious a problem for our interviewees as we first believed, but failure to enjoy what is read, i.e., the assigned textbooks, did contribute to their alienation from reading.

Our informants' descriptions of their attitudes about school subject matters, as we saw in the previous chapter, focused soley on teachers and teaching skills. Farrell (1990) corroborates this finding with his study of New York City school leavers: 'students distinguish "good" from "boring" classes on the basis of the process rather than the content of teaching' (p. 108). A Pittsburgh female student liked her biology course because of its tactile emphasis: 'Because you got into it. It wasn't something that they sat up there and taught. They actually let you cut up frogs and things like that'. These narrators did not complain about the inherent difficulty of a subject area. Subject mastery and preference rested solely on the teacher and teaching. Another female narrator describing another biology class and instructor liked this subject, and pointed directly to the teacher's approach to pedagogy and caring:

> What we did. His projects. It was hard, but he made everything look easy. He would explain it real carefully. He would sit down and talk to you. If you didn't know it, he would sit down and talk to you. If he had to show you, he would show you. He would talk to you first. We watched movies and dissected.

An African-American female student liked her math class also because of the instructor: 'I just found math easier than other subjects. I had a nice teacher. She took time. You could come before or after school. She would stay when it was convenient for you. She would always be there'. Science and mathematics are usually the bane of many marginal students. Yet, these school leavers, in a seemingly atypical fashion, liked them. Good pedagogy and caring made the difference, as we can see in this description by a female narrator who intensely disliked her high school history class and contrasted the dreary teaching methods in that class with her middle school experiences:

> They were talking about dead people (laughs). Dead people — I thought we should be talking about what goes on around at that time, because it would have been more interesting. We could all relate to what was going on. I couldn't get into it. The only way I got interested in it was that one year — eighth grade. We used to play history games, boys against girls. But after that we got up to high school. In high school, they would give you the book, ditto sheets, and give the answer questions, and give you a test on it. I mean that's all that they were doing.

Only a few informants described their preferences for certain subjects regardless of the teacher and pedagogy. They exhibited curiosity about subjects they liked. One African-American male's favorite subject was reading: 'I like to know what is going on. I would put my mind in that'. He liked English because it involved reading. An African-American female preferred English and history because she liked to read and learn about the past. She also made positive comments about algebra and Spanish. She too enjoyed social studies, especially a focus on law, since she wanted to become an attorney. She spent so much time completing her homework, often toiling until two o'clock in the morning, that she never participated in extracurricular activities. She abandoned her schooling at age 16 in order to earn a GED and start college a year sooner, at age 17. Leaving school, in this case, represented a 'shortcut' to higher education. Finally, many students fondly recalled taking field trips to Pittsburgh's Buhl Planetarium, Heinz Hall for the Performing Arts, and Carnegie Museum and Science Center, and to the Hershey chocolate factory in the eastern part of the state. Most expressed enthusiasm for these learning experiences, which broke the monotony of the classroom routine.

Another sharp criticism we heard stemmed from rigid scheduling and subject redundancy. One interviewee described the worst thing about school: 'Getting up! School should (start) later or something. That's why I can't wait until I start college'. That student preferred flexible schedules, which she believed existed in higher education. Also, the repetition of subject matter content bored many students, as reflected by this narrator's comments:

> I don't think you should need all of those years in school, because what they was teaching me in tenth grade I had already picked up from the eighth grade. They want you in school twelve years, thirteen if you go to kindergarten. I don't think you need all them years of school.

This student had never repeated a grade, yet the subject matter seemed to her to be the same year after year. Still another informant abandoned high school because she felt intellectually unchallenged: 'They was giving me the fifth-grade work'. The schooling process proved to be valueless to her: 'I didn't like it. It wasn't doing nothing for me. I was just going over the same stuff I already knew'.

The Extracurriculum

School leavers rarely participate in co-curricular activities, particularly athletics (Ekstrom, *et al*, 1986, p. 36; Kelly, 1993, p. 207). This reflects alienation from the schooling process. Kelly (1993, p. 207) found that many such students likened involvement in 'voluntary school activities, defined by adults and dominated by high-status peers', to 'participating in their own oppression'. Boys generally perceived 'most extracurricular activities feminine', whereas 'girls derived more status, particularly among girls, from participating in extracurricular activities' (pp. 207–8). Most of our narrators too avoided these informal educational activities, listing a variety of reasons. One female student feared peer ridicule: 'I guess it is because when I do something and somebody laughs at me, I get very upset. . . . They do laugh at you, not cause something is wrong with you, they just laugh. I don't like that'.

Some exceptions existed, though. A few of these students participated in co-curricular activities, like this male informant: 'I liked the extracurricular activities — the band. I loved the saxophone. I loved the band. Something that made me a part of something else — the chorus. I was part of a group; a collective group and we did things. We went places'. A female narrator, very active in non-academic activities, expressed similar enthusiasm: 'After I got into high school, I was on the marching band, the orchestra, and then all city. I stayed in music. I wasn't interested in anything else'.

These narrators actively participated in the co-curriculum, yet abandoned school. The extracurriculum can compete with, rather than complement, the formal curriculum, contrary to early Progressive beliefs. More and more African-American students appear to be remaining in school, but their academic progress has not reflected this trend. 'Researchers studying the schooling process of Black working-class students have concluded that these students commit themselves to a non-academic sports subculture that limits their academic participation' (Solomon, 1989, p. 80). Thus, staying in school while disengaging from academics results in a phenomenon labeled as 'in-school dropouts' (*ibid*, p. 79). This is significant because when sports is no longer available, these students complete the academic disengagement process by abandoning school altogether. Solomon (*ibid*, pp. 80 and 84) summarizes the impact of sports culture on academic studies for African-American students:

> First, it tends to interfere with their academic progress: students miss classes, shortcut homework, and do poorly on exams. Second, Black students who are preoccupied with sports are excluded from other areas of school life, especially those that are academically oriented, such as speech, drama, or student

government. Finally, Black sports culture negatively influences intergroup relations; Black control of sports, coupled with White withdrawal, contributes to divisiveness within the social system of the school.

In this case, the informal curriculum actually narrows African-American opportunities rather than expands them.

Other informants dismissed the extracurriculum outright. A White female saw these after-school activities as superfluous: 'They were all right, but I could have learned more if the teachers would have showed us more. I could have learned more there than I have anywhere else'. According to this student, the formal curriculum is more important than the informal one. However, the formal curriculum proved to be largely ineffective, as we have seen.

Work

In spite of these generally negative perceptions of school content, both formal and informal, 85 per cent of these students valued schooling over work. One male student summarized it as such: 'As I look back on it, school was more valuable than work because work is important, but it can only take you so far'. Another informant echoed his comments. She preferred work over school: ''cause I didn't have to sit in them hot classrooms'. Yet, like the overwhelming majority of the students we interviewed, she saw schooling as more important in the long run: 'Because you need an education wherever you go. You are not going to always find a job without an education'. These narrators' comments about the relationship between work and schooling totally contrasted with the perceptions and experiences of their nineteenth- and early twentieth-century counterparts, reflecting profound changes in the political economy.

Some qualified this opinion, however. Work held value and 80 per cent of the students preferred it to school, though not out of some abstract notion, but for its monetary rewards. One African-American male informant worked in building maintenance while still enrolled in high school. He liked work because it provided him with money for clothes. Nevertheless, work interfered with school.

Mostly, like homework, I guess . . . When you had to clean the building, you had to be there. The building would have to be clean by the next day. If I said I would do it, I had to do it. It wasn't like I could call off. Sometimes I may be real tired, cleaning up. I would do some of my homework, or none of it maybe. I would fall asleep. Maybe I would get a zero. I didn't like getting zeros. I wouldn't go (to school). I would stay home and do my homework and turn it in the next day, just to keep from getting a zero.

In addition to financial incentives, all of these narrators appeared to possess a strong work ethic. When an interviewer asked an African-American female what she liked to do for 'fun', she responded: 'I like to work a lot. I have fun working, 'cause I got friends there and do something to make money'. A White female student, who had worked since she was 15 years old, commented: 'I know it sounds weird, but if I don't work I would go nuts'. She, like most of the narrators, preferred work to schooling. Unlike many narrators, she valued work more than schooling. 'I think that work is valuable 'cause it gets you money; it gets you what you gotta get. You can take care of yourself'.

Although most of our narrators placed a higher value on schooling than work, they never expressed a love for schooling, unlike the 'second chancers' Kelly (1993, p. 202) found and interviewed in her study of California's continuation schools. They 'liked academic learning'.

This group valued the chance to come from behind and responded enthusiastically to the incentives the continuation school held out. . . . Second chancers were intellectually capable and fairly well prepared but had got derailed from school, usually because of a personal or family crisis, rebellion against the nonacademic aspects of schooling (student cliques, petty school rules), or both.

Pittsburgh dropbacks resumed schooling to secure a job or improve their work opportunities, in contrast to Kelly's (1993) second chancers who mentioned 'increased maturity or "real life jolts"' (p. 202), or crises, prompting their school return.

Our informants' perceptions about the importance of education in helping one secure gainful employment appears to be tragically accurate. School leavers, between 16 and 24 years of age, face an exceptionally high unemployment rate of 25 per cent. They compete not only with each other, but also with high school graduates. Markey (1988, pp. 39 and 40) breaks the figures down by gender and race to reveal an even worse situation. Labor force participation for female school leavers appears to be especially poor. 'In October 1986, the jobless rate for female dropouts age 16 to 24 was 30.4 per cent, about two-and-one-half times the rate for women this age who had ended their education with a high school degree'. African-American school leavers face an unemployment problem 'approaching crisis proportions'. They suffer a 44 per cent jobless rate compared to 18 per cent for Whites (*ibid*, p. 40). To exacerbate matters, many school leavers find themselves unable to even gain any work experience, which could enhance their employment possibilities. 'During 1986, 17 per cent of men age 20 to 24 with less than four years of high school had no work experience at all; 25 per cent had worked 26 weeks or less' (*ibid*, p. 41; see also Weidman and Friedmann, 1984, pp. 27 and 28). The school leavers we interviewed appeared to be acutely aware of this relationship between work experience and job opportunity. As one African-American male stated it: 'I worked to make something out of myself — get credit. Like if something comes up in the future, you can say he did this and he did that. . . . I ain't lazy. I like that'. Finally, some of our informants, following historical tradition, simply worked to help support their families.

Regardless of the reasons, working contributes to school abandonment. 'Research has found that working more than ten hours per week places adolescents at a higher risk of early school leaving even when controlling for family background, school performance, measured achievement, and other background characteristics' (Roderick, 1993, p. 31). Working detracts from school responsibilities; it takes time away from homework assignments, saps energy necessary to concentrate on studies during the schoolday, and offers a concrete cash reward rather than an abstract promise of future success.

Graduate Equivalency Diploma

All of the school leavers we interviewed were actively pursuing a Graduate Equivalency Diploma (GED). 'In 1983, 740,000 persons sat for the examination. They reported an average of 9.8 years of schooling, and a full 33 per cent were age 19 or younger' (Fine, 1991, p. 86). For some researchers, however, the value of this alternative route to a high school diploma appears to be in doubt. Roderick (1993), for example, questions the GED's worth, referring to a study which 'found that labor market outcomes for males who earned GEDs were much more similar to those of high school dropouts than to those of high school graduates' (p. 3).

The literature seems to be less definitive, though. On the one hand, Kelly (1993) cites studies that question the equality of the traditional and non-traditional approaches. On the other hand, other information muddies the water: 'Anecdotal evidence suggests that employers may not distinguish between . . . . (the) diplomas, but military recruiters do, arguing that alternative-program graduates are twice as likely to leave the service because of drug and disciplinary problems' (*ibid*, p. 85). One of our female informants also saw ambiguity here and took the analysis one step further. She questioned the value of both the high school diploma and GED; they merely get you a job. Rather, a college degree represents the only bridge to decent work: 'Nowadays, (employers) are going by college. . . . . They are not just going by no diploma no more'. She saw the GED as a bridge to more education, and ultimately to a career. Finally, Powell, Farrar and Cohen (1985, p. 42) maintain that the GED cheapens the high school diploma because clever students can abandon school early to pass the equivalency examination and begin community college sooner. Two of our narrators claimed to have chosen that strategy, that is, leaving high school when they were 16 years old in order to begin community college at age 17.

What many of these studies ignore is the affective side of completing a 'degree'. Our narrators, whether naive or uninformed, reflected a positive attitude toward the GED. They often appeared proud of their ability to earn an alternative diploma. An African-American dropback, after a long bout of unemployment, identified school with work: 'School is just like work'. She elaborated the relationship that she, and so many other informants, saw between school and work:

> If you don't have your diploma or GED, how are you going
> to get a job? I would say that you have to go to school first.
> A lot of people work with no diploma, but that's not making
> no money. Everybody wants to make money in life — a lot of
> money, not minimum wage, like $10 or $15 an hour or even
> more.

Our narrators appeared to be consistent in describing this relationship. Did this reflect their own value systems? Or did this represent the product of Job Corps instruction? With so many informants interviewed over such a long period of time giving such a uniform response, we tend to think that this represents the Job Corps's influence, which we treat in chapter 10. When the interviewer asked this same dropback how she felt about her classmates and friends graduating from high school and receiving a diploma, she responded that she 'was glad they made it'. She continued: 'They were strong enough to hang in there and do what they have to do'. This implies, of course, that she was weak. However, this in no way demeaned her Job Corps education or devalued her GED; she appeared, in fact, to be effusive about her Job Corps experiences: 'Here, I think it's a better opportunity. The GED, I think, is a little bit higher than a high school diploma. I think you have to work a little bit harder'.

The Social Curriculum

Finally, and briefly, our narrators, throughout the interviews, alluded to another source of knowledge. Few acknowledged it, and even fewer described it in any detail. Nevertheless, through their interactions inside as well as outside of their classrooms, students clearly learned from each other. As one African-American female expressed it: 'You learnt. You can't just be school educated. You got to be street educated too. 'Cause if you're not, you can't get through it; 'cause people can run a game on you fast. So, I learnt from the street. I learnt a lot from the streets'. Accordingly, one needs to recognize that education, not mere schooling, includes a totality of experiences which promote learning. How the school relates to and builds upon such informal education constitutes an important building block for the schooling experience.

Conclusions

Our informants painted a complex and perceptive picture of what they learned, eroding many stereotypes. The formal curriculum, which they saw as based on limited and irrelevant materials and dreary and uncreative pedagogy, seldom captured the interests of potential school leavers. However, most narrators read informally, all enjoyed what the stimulating and committed instructors taught, and everyone spoke positively about their visits to science, art, and production centers, i.e., the real world. The extracurriculum, for a variety of reasons, largely failed to stimulate and hold these students in school. They rarely participated in it because they either found it worthless or, from a practical standpoint, were simply not there because of absenteeism; others went to work. The ultimate irony was that most informants expressed a preference for work over schooling, but an overwhelming majority saw schooling as more valuable. This latter perception stemmed directly from their job experiences, which proved to be low-level, ill-paid, and dead-end. Schooling represented the only bridge to worthwhile work for them. Job experience had taught them a hard lesson; that led them to choose the GED path and enter a job training program. This school resumption process will be treated in chapter 7.

Why School Leavers Left and Why They Returned

I had the baby. I wanted to take care of her.

I got kicked out. I didn't drop out of school. (Pittsburgh school leavers)

We argue that patterns of school leaving are complex, often enigmatic, and we recognize distinctions among dropouts, pushouts, fadeouts, and easeouts among our narrators; combinations of these categories likewise exist. This pivotal chapter describes and analyzes these various processes of disengagement and leaving school. It treats each category through examples from our narrators, and each section includes an in-depth illustration, recounting in detail how they came to leave school. We also reconstruct the decision to return to school, and conclude this chapter with an in-depth analysis.

Bickel, Bond, and LeMahieu (1986, p. 3) question whether there is a single definition of a dropout. They use the phrase 'early school exit' (ESE) because it appears 'more comprehensive and thus a more accurate means of characterizing this complex issue'. They note 'that discussions of the "dropout" question too often have had a narrow view of the problem' and conclude that students who left school before completion fall into at least three categories. First, the *dropout* consciously decides to leave school early for a variety of reasons (for example, disciplinary problems, low achievement, pregnancy, etc.). Second, the *pushout*, rightfully or not, perceives the school and/or its personnel as hostile. Third, is the *fadeout* whose decision to leave school does not occur at a particular time and is a less conscious choice. A series of events over time typically embody characteristics of the first two categories and gradually build up. The fadeout student usually finds it difficult to point to any one event or the exact time when dropping out became a reality. We add a fourth category: the *easeout*, who abandons schooling with either administrative or teacher 'encouragement', or both. 'Because this can never be an official policy, (such encouragement) is conducted sub rosa and, for good or ill, it is a reaction to intolerable situations, intolerable for both students and teachers' (Farrell, 1990, p. 95).

Social forces, like race and class, also play important roles in the school leaving phenomenon, as we have seen. Gender offers further complexity. 'Among working-class and low-income students across racial and ethnic groups', Kelly (1993) found, 'girls and boys often disengage for different reasons, in different ways, and with different consequences' (p. xvi). Girls, on the one hand, tend to see their school disengagement as personal, or, as Kelly phrases it, 'the soap opera analogy' (*ibid*). They tend to emphasize 'relationships' with teachers, family members, peers, and boyfriends (see also Fine and Zane, 1989, pp. 28–9).

> In daytime television serials, early pregnancy, illness, divorce, criminal activity, drug abuse, affairs, domestic violence — problems girls on the margin experience in their own lives and at school — are prime subject matters. Many educators consider such issues 'personal', unfortunate distractions from the formal curriculum. The high schools' failure to acknowledge and make connections to students' soap opera-like experiences contributes to the disengagement process.

Boys, on the other hand, usually describe conflict: 'Disengaging boys are more likely to compare high school to prison where principals are wardens and teachers, guards; these boys often cope with their mandatory confinement by flaunting the rules and fighting with teachers and peers — a direct action approach more likely featured in cop shows than soap operas' (*ibid*, pp. xviii and 94–95).

Kelly's use of a gender lens to examine school leaving is significant because the differences in the experiences of males and females have been historically ignored. Boys have received more attention from usually male researchers, Kelly theorizes, because they see boys as 'future breadwinners' (*ibid*, p. 6), that is, 'a high male dropout rate suggested a potential shortage of productive labor'. This human capital mentality has driven the ongoing school reform movement. Female labor has not received as much attention, because, according to gender stereotypes, girls are destined to assume traditional domestic roles. Implicit in all this is that female labor remains devalued. Finally, Kelly writes, scholars have tended to conceptualize the reasons for dropping out as school-related rather than personal or family-related. In doing so, they have inadvertently directed attention away from the interaction between institutions. Do girls leave school because of early pregnancy, or is early pregnancy a means of escape from an institution — the school — that has failed to offer them a sense of purpose and competence? Researchers have implicitly assumed that school has had little influence on girls' decisions to leave early to get married or have a baby. (*ibid*, p. 6)

Kelly, unlike Bickel, Bond, and LeMahieu, avoids using the term *dropout*, as well as the label *pushout*. *Dropout* blames the individual while *pushout* indicts the institution. More important, these terms describe a 'final outcome' rather than a cumulative process. 'For a significant number of others, the decisive moment of dropping out or being pushed out never occurs; these students attend infrequently, leaving and returning several times, and thus may be more aptly described as fade outs'. Kelly (*ibid*, p. 29) prefers to view 'early school leaving' as a 'mutual process of rejection, or what may be called disengagement'. This unfolds over time and consequently appears 'difficult to document'. Kelly (*ibid*, p. 30) points to four experiences leading to disengagement. First, *academics* encompasses poor achievement, remediation, grade repeating, and suspension and expulsion. Second, *peer relations* involves fights, isolation, and association with similarly, marginal students. Third, school disengagement also means nonparticipation in extracurricular activities. Fourth, when the diploma seems unnecessary, insufficient, or unattainable, schooling becomes empty and students devalue it. Kelly thus presents an unfolding school leaving process, while Bickel, Bond, and LeMahieu see it in a more conventional way, as a culmination. We emphasize both process as well as the act of abandoning school.

Dropouts

We interviewed sixty-four narrators who we categorized as dropouts. Their descriptions of the school setting and their perceptions of school personnel indicate that, many of them became progressively marginal in school, and a single event often triggered their departure. For one male student, whom we labeled the 'lost soul', his tenuous relation to school depended solely on athletics. And when this abruptly ended, his fragile world simply collapsed:

> My gym teacher caught me smoking a joint. That was the only time I got suspended. It was a hurting blow, 'cause I got kicked off the football team. Everything just went downhill after that. Yeah. We don't need no druggies on the football team. That jock stuff. Well, forget it. I started to get high even more. Then I started cutting or leaving school early. That led to just stop coming period. Just stop coming. I wanted to know what was going on in the streets so bad. I regret that now.

In other cases, a variety of factors led to disengagement. Peers, for some students, represented the primary influence. As one narrator explained: 'Streets, peers, peer pressures; they was cutting, so I would cut'. A few informants described how their parents incurred and paid fines for their ditching school. A White female recalled the morning routine and dialogue:

> I used to tell my mother, 'Yeah, I'm going to school today, Mom'. And she would say, 'Are you sure?' I would say, 'Yeah'. She thought I was going to school every day. She had to go to court 'cause I didn't go to school and everything . . . Just a small fine and (I was) sent back to school.

Another narrator, a White male, accumulated $1600 in fines for missing school. Parental authority played another role, occasionally with unanticipated outcomes. This youth explained 'I had a lot of pressure on me. My family didn't think I was doing the best I can. I think I was putting quite a bit of effort into it. They don't think I was doing enough. I tried to keep an average. I never flunked'. Others just ran away from home and from school. In some cases, students abandoned school as soon as they reached their sixteenth birthday; this occurred with one ninth grader. Still others gave social and idiosyncratic reasons for leaving school. One male student was self-conscious about wearing glasses. His father purchased contact lenses, but when he lost them he quit school rather than face social humiliation. Finally, some just could not face another day of school, as in the case with one female narrator, who expressed disdain for the entire schooling process: 'The worst thing about school . . . another day. The worst thing I had to face was going to class, because I knew I wasn't gonna learn nothing. I knew that it was just another day like any other days.'

Pregnancy

Twenty-eight, or 48 per cent, of our fifty-nine female informants left school because they became pregnant. Some of these we labeled as dropouts, while others we categorized as easeouts. We based this distinction on the school leaving process, that is, we attached the former label to women who abandoned school because of pregnancy, and the latter to pregnant students who were referred by counselors and other school officials to Letsche Education Center, the district's continuation school, where there is a program (Ed Med) for pregnant students. Twenty-four, or 49 per cent, of African-American female dropbacks and four, or 40 per cent, of female Whites were pregnant in our sample. Of the twenty-five who told us the number of children they had, nineteen had one child; the remainder mentioned two children. Our female narrators averaged 18.81 years old, with the overall sample mean at 18.56. African-American females appeared to be slightly older, at 18.88 years while White females averaged at 18.5 years of age. They all gave birth very young. The youngest dropped out when she was attending middle school. She simply stated, 'In seventh grade I got pregnant'. In general, pregnancy proved to be a difficult and frightening time for her, as well as for others.

> If I hadn't gotten pregnant, I would still be in school. I felt uncomfortable. It was my first pregnancy and I didn't know what to look forward to. I didn't have anyone to talk to. I just stayed more confined to myself. When I was pregnant, during the whole nine months, I stayed sick all morning long. If I had stayed in school, I would have been throwing up and uncomfortable and miserable . . . I lost a lot of weight . . . I stayed depressed a lot. I didn't know what it was like having a baby.

Pregnancy alone did not contribute to dropping out. The lack of child care created serious problems for these young women. The Pittsburgh schools offer a limited child care service, but high demand soon overwhelmed its capacity. As one woman related: 'I signed up my son for that, but he was on a waiting list'. This forced her to remain home to care for her baby rather than attend school. The key point here is that she, and other young mothers like her, wanted to resume their schooling, but could not because of inadequate support systems. Other women felt obligated to withdraw from school to support their children. One informant described her new, serious responsibilities: 'I had the baby. I wanted to take care of her. I wanted to go out and get a job and support her like a regular person . . . I don't understand how you can be 16 years old and pregnant' (and remain in school). In all cases, these women recounted a difficult struggle to ensure their own and their children's survival while continuing their schooling:

> I don't see any girl staying in school if she is pregnant. I know a couple of my friends that did stay in school and did stick it out. They did graduate and they do go to college and they still have a kid. That's what really makes me mad, that I didn't do the same; different circumstances, different people.

Teen pregnancy represents a serious problem, but politicians and the media have sensationalized this phenomenon, somewhat exaggerating and distorting it. Adolescent pregnancy and childbearing peaked in this country in 1957, and actually declined by 42 per cent between 1955 and 1986 (Jones *et al*, 1986, p. 37; Vinovskis, 1986, p. 161). However, two variables have changed. First, and most important for the schools, 'pregnant teenagers have become more visible to their schoolmates, teachers, and the general community since, in 1972, it became illegal to expel a student from school because of pregnancy' (Jones *et al*, 1986, pp. 37–8). While pregnant teens are no longer 'pushed out' of school, girls still tend to drop out because of a lack of an effective and sensitive support system. Second, fewer pregnant adolescents marry. Out-of-wedlock teen births, to mothers age 15–19, have risen dramatically during recent decades, 'from 72,800 in 1955 to 280,300 in 1985 — a 285 per cent increase in three decades'. Serious racial differences exist as well: 'While 45.1 per cent of White teen births are out-of-wedlock, 90 per cent of Black teen births are out-of-wedlock' (Chase-Lansdale and Vinovskis, 1993, pp. 204–5; see also Jones *et al*, p. 38; Vinovskis, 1986, p. 161).

Two other issues, according to the Guttmacher Institute's thorough study of teen pregnancy, have added some ambiguity. The increased use of abortion since 1973 has somewhat distorted these figures. Teenage pregnancy rates have remained stable, while birthrates have dropped. 'The percentage of adolescent pregnancies carried to term dropped from 72 per cent in 1973 to 55 per cent in 1979 and has remained stable since that time' (Jones *et al*, 1986, p. 40). Social class differences, perhaps superseding race, also remain obscure. 'Although race is sometimes used as a proxy when data by socioeconomic status are not available, it is not clear to what extent differences in adolescent pregnancy between the two racial groups are due to income or to other factors'. We do know that in 1981 the federal government declared 41 per cent of African-American and 14 per cent of White teenage females as poor (*ibid*, p. 43). A clear and sober picture of adolescent pregnancy continues to elude us. Until this society and its schools come to grips with this situation, too many overwhelmed and fearful young women will continue to feel lost and alone.

A Personal Perspective

Personal crises directly correlated to leaving school. An articulate and sensitive African-American male, related his dropout decision to his home problems and a mother ill with cancer. This responsibility dominated his social and academic life. Except for his basketball teammates, he had few friends: 'I had to get home and take care of my mom. I stayed with my mom a lot. If the home nurse wasn't there, I would have to be there. I couldn't do too many things'. The visiting nurse's irregular appearances directly affected his school attendance, forcing him to make a painful choice: 'Because of my attendance . . . it was my mom. It was more what do I consider: staying with mom or going to school. I figured I would rather stay home with my mom. . . . I was considered the responsible person'.

Some of his school's staff offered support: 'I was always close to my vice-principal. If I had problems, I would see him. When I was getting ready to leave (school), I talked to the vice-principal. He wanted me to stay'. However, because this student had missed so many days of school, he had to repeat a grade. He had already repeated ninth grade. He continued, 'I had problems with my mother. I would call my vice-principal and tell him why I was staying home'. The secretaries too offered help: 'I was known around the school. The secretaries knew me and got along with me. If I had problems, I would talk. If my mom called, they would call me into the office right away and tell me to go home. The secretaries were all right'. A home tutor, sent by the school, could have resolved his school absences.

Pushouts

Pushouts appear to be a national phenomenon. In Maryland, 'a fourth of the dropouts studied have been suspended from school at least once and that an additional fifth have been determined by teachers to be classroom problems' (Beck and Muia, 1980, p. 69). A study of Arizona and New Mexico school leavers, by Stoughton and Grady (1978) pointed to typical reasons: 'non-attendance, lack of interest, and disciplinary difficulties' (p. 315). However, these researchers concluded that the 'frequency with which dropouts fell into these three categories suggests that schools may be creating "pushouts"'. They then raised a profound question: 'Are these categories administrative expediencies for labelling, rather than explaining why the student drops out? If this is so, then one may be led to question the validity of the concept of compulsory education'.

That statement raises a provocative issue: on the one hand, how long (can, or should), public school administrators tolerate and attempt to ameliorate what they regard as deviant behavior? On the other hand, do school officials stereotype students who exhibit deviant behavior as 'losers' and thus subtly contribute to such students leaving school? We recognize that school administrators, face a dilemma. It should be emphasized that we designed our research to identify *why* students left school before completion. Our findings, consequently, focus on students' perceptions of school and why they decided to leave, and, as such, are intended to enlighten administrators and teachers about a persistent and complex problem they face everyday with few solutions.

Eleven of our informants did simplistically describe a pushing out process; ten were African Americans, equally divided between male and female. The lone White male student did not mince words about his experience: 'I got kicked out of school'. However, school authorities gave him an option: 'They sent me to night school and if I passed two classes in night school, I could come back over here'. He too left that program for too many absences: 'I didn't get transportation. They don't give it to you when you go to night school. That's on your own'.

School district procedures on suspensions and expulsions sheds a different light on this example. Being 'kicked out', to use the narrator's words, could mean either suspension or expulsion. His comment that he was given an option implies that his infraction led to a suspension that could be reversed by successful completion of an alternative educational program. Was he 'kicked out' in the sense of being expelled permanently? It is fairest to both the student and the school district to infer from his statement that he equated being suspended with being kicked out. Was the condition for reinstatement adequate in his situation? Apparently not, for he saw suspension the final episode in a series of incidents; he interpreted this as being pushed out. However, as the informant noted, district officials stipulated that he attend night classes; this occurs only in cases of excessive violations of the school codes.

Sometimes there was only a fine line, for some of our informants, between being pushed out and actively dropping out. In some cases, when vice-principals proceeded to fine families when students skipped school, it provoked them to completely terminate their schooling. An African-American female abandoned school, or 'signed out', because she did not want her mother to be harassed: 'People's mothers would be going to jail. I believed that was why I quit school. I couldn't put up with that'. She elaborated:

> If I wouldn't have signed out, they would have kept sending my mother all those fines and they would have kept trying to put my mother in jail. My mother would get us up for school and it ain't like she can't follow us to school. I was like seventeen, and she thinks we would be in school. We do go to school, and we would be cutting classes and have a little bit of fun and everything. It wasn't really my mother's fault. She didn't know I was cutting classes and stuff.

This fining system operated as a vestige from the early decades of the century, when the district used compulsory attendance officers who punished students and parents alike.

Gender and race appear to be directly related to pushout rates. Nationwide data from the National Center for Educational Statistics (1992, p. 31) indicate that more males, 19.2 per cent, than females, 12.7 per cent, experience suspensions. Expulsion figures reflect a similar trend: 13.4 and 8.9 per cent, respectively. However, the most glaring discrepancy occurs with race, with African-American students suffering the highest pushout rates. In 1990, 26.3 per cent nationwide claimed that they were suspended while 24.4 per cent were expelled. These numbers sharply contrasted with 14.5 and 12.5 for Hispanics and 13.1 and 8.7 for Whites.

Similar patterns exist for Pittsburgh students. District officials rarely, if ever, expel students. However, suspension records between 1977 and 1984 illustrate that males averaged 72.4 per cent of the total students suspended, with the remaining 27.6 per cent, being female. African-American suspensions during that same period amounted to 77.4 per cent of the total; Whites compiled 22.4 per cent of the total. the district later switched to a different data reporting system, and reported that 21.3 per cent of the total student body experienced suspensions during the 1992/93 school year. Racial differentiations remained, because 30.1 per cent of African-American students faced suspensions that same year compared to 13.1 per cent for 'other students'.

Between 1979 and 1984, the three leading reasons for suspensions included, in order: 'disruption of school', 'repeated school violations', and 'physical abuse of a student'.[1]

A Personal Perspective

The pushing out process, according to one African-American female who was, expelled in ninth grade, follows a gradual, but steady, escalation in authority conflict. She repeated ninth grade three times: 'When I was in my first year of (high school), I was good. I hardly got sent (to the vice-principal's office). Then, my second year I did poorly. I used to go to the office a lot'.

School authorities pushed her out during her third attempt at ninth grade over a locker incident. She had been sent to the office on other occasions because she failed to bring her required books to class: 'I always had a locker that always got jammed . . . I keep going to them and telling them about my locker, and they still won't transfer me to a different locker. If I shared something with somebody else, they would say you can't do that'. The vice-principal ordered her to carry all of her books with her. However, this was impossible because they remained locked in her jammed locker: 'That was one reason I got suspended. I got in trouble just 'cause I couldn't get my own books out of my own locker. They wouldn't take the lock off'.

Her conflicts grew more intense with another building vice-principal, ultimately resulting in physical confrontation:

> We was like arguing, and she was telling me I got three days (suspension) unless my mom came up. I turned my back on her. She grabbed me and turned me around. You can't put your hands on me. It ended up in she hitting me and me hitting her back. She told me not to come next year.

Fadeouts

The fading out process has many subtleties and complexities. For these reasons, we only placed five of our informants in this category — all African-Americans, four males and one female. The female student had walked to her new high school. She had transferred there because of a residence move at the end of the ninth grade. In contrast to her previous year's experience, she found her tenth-grade environment unpleasant. Neither her teachers nor her peers befriended her. She recalled no opportunities to establish meaningful contacts; that was her reality. The school was large, but clean and orderly. It enrolled well over 1200 students. The principal at that time appeared efficient and competent, yet somewhat invisible. This school was uninviting and unfriendly to this student, almost a hostile place. She increasingly dreaded each day she walked to her new school. Then, one morning she just walked past it and found that act very easy, and she never returned.

A Personal Perspective

The clearest single illustration of fading out occurred with an African-American male informant, who had transferred from an affluent suburban school district to an inner-city school in Pittsburgh during his senior year. This change traumatized this articulate and thoughtful narrator: 'My experience at (a city high school) is probably what made me drop out of school, because I came from a suburban school instead of a public (*sic*) school'. His former school seemed so different that throughout his interview he consistently failed to recognize that it too was a public school. He described a gradual disengagement process, consciously unplugging himself from peers, teachers, and the school's staff; his classes bored him in particular, and the school alienated him in general. This experience exacted its toll, because in January of his senior year, after having regularly skipped school, he simply stopped attending altogether: 'I dropped out of school because of culture shock. I could not relate to the change that I had made . . . And then I dropped out because I had money. I had money and I was earning it honestly. I felt like I was taking care of myself, and I could continue to do this without going to school and having to deal with the problems that I was facing in school'.

The students at his new school presented him with his first real change:

> I just could not relate to them. I knew a lot of them . . . but as far as at school activities, being around a group of people, I stayed away. One, it seemed like everybody was into being disrespectful to the teachers; everybody was looking for a way to get over her some kind of way. Then you had some students who were interested in graduating, but it was basically just take all the easy classes, just to get out.

His comments show that he abhorred the former group and felt contempt for the latter one; no one seemed to be academically engaged. Poor student behavior, for this narrator, justified the presence of a school security force: 'If there weren't any security officers, the students would probably run the school'. However, unruly student conduct paled when compared to what he termed student violence: 'The students were sort of violent; they were very violent. I had never (previously) heard of being able to curse out a teacher. I had never heard of hitting a teacher and jumping on a teacher. I had never heard of that. On a daily basis, they would get up and walk out of class, and don't come back'.

Cynical and calloused teachers reinforced his alienation: 'To me the difference as far as the teachers was concerned, was that they sort of had an "I don't care" attitude. Get you in and get you out. Just so long as you get that D'. These teachers also cultivated favorites, usually, but not always, based on behavior. 'The type of student that would be their favorite would be good students . . . that didn't cut up in classes. It didn't have anything to do with grades'. This narrator only pointed to one exception, a music instructor: 'He was a good teacher, because he was very concerned about what you were doing. He wanted to make sure that you were doing good in all your classes. He was good. He knew that I was having problems, and he tried to encourage me'.

Except for physics, which 'was a challenge', he thoroughly disliked his courses. This related directly to obstreperous peers and a drudgelike classroom routine:

> On a daily basis, one of my classes, everyone would come in and it would take the teacher twenty minutes to get everybody quiet. And finally they all would start doing their work. You weren't taught anything . . . You went step-by-step with the book. They had like booklets that you would do. By the middle of the class all of a sudden somebody would try to distract the class. Then he or she would get into it and they would curse out the teacher and walk out. And then by that time everybody is laughing and talking and the class is disrupted. And then you just get everyone not doing anything but talking and socializing. By that time, it's time to go.

All of his comments touched on the confusion, disarray, and noise typifying that school's environment.

School counselors offered him little solace, only superficially addressing his class schedule.

> When I first enrolled in the school, my mom agrees, it was like a rush, rush deal. (The counselor) wasn't really concerned about the subjects I had taken before. She said that I had come from a suburban school and I would probably do good. And then I had to go back to her and have some of the classes changed. She wasn't concerned about getting my records. She said, 'I will put you in these classes until your records come'. I had to go to her to see if my records had come. I didn't have a good rapport with any of the counselors.

This lack of interaction sharply contrasted with his former experience: 'Our counselors from where I came from were very personal. They would call at home, from their home and discuss it. And then I came here, to the (urban) public school'.

Finally, subtle as well as blatant racism confronted this student, which further eroded his connections to school. His recollection of racism emerged indirectly from a question about field trips; he pointed to a trip he and fifteen classmates had taken to Howard and Tuskegee universities on a charter bus. He observed that they only visited those two schools ''cause all the students were Afro-American'. He disagreed with this strategy: 'I felt that we were in (Washington) D.C., and there were like George Washington University, Georgetown, and American universities. There was like three or four other universities that we could have seen, but we didn't. We just went to Howard and we just went to Tuskegee'. He then described the selection process by which the students voted to visit these particular institutions: 'There was a meeting and the students decided where they wanted to go. But, of course, all the students want to go to Howard, because everybody knows about Howard. They hear about how much fun it is. I felt it shouldn't have been up to the students, because of a lot of things we don't know'. This narrator strongly objected to this procedure: It disturbed him that neither teachers nor counselors, who served as sponsors and chaperones, offered any other options to these African-American students except traditionally African-American institutions.

While this racist incident appeared somewhat ambiguous and is open to various interpretations, this student recalled more obvious episodes in which he felt slighted because of his race:

> In my physics class, we were doing an experiment, and I was having problems with my lab partner 'cause my partner was

never there. I went to the teacher and asked him a question. Another student was in front asking a question also. He helped that student, and I went to ask my question again and he like skipped over me and went to this White girl. She was really smart and she always did her work. He like helped her . . . I felt like, well I was doing the best that I can being that I was by myself . . . That one teacher I felt it all the time.

He then continued to recount other racist incidents: 'A lot of times with other teachers it was more subtle. Like if they would see you in the hall they would ask you for a pass. If they would see a White student walking past they might say, "Keep on walking"'.

This high school, according to this fadeout student, lacked a nurturing environment, and therefore did not represent a good place to learn. The cumulative effect of these experiences proved overwhelming to him: 'You don't have teachers that are concerned about you. You don't have students that are concerned about themselves. You don't have students that are respectful to the teacher'. Attracted by a pleasant job as a salesperson, this student just faded from the scene.

Easeouts

Easing out represents a surreptitious form of leaving school. This process involves the approval of teachers and administrators, if not their encouragement, to leave school. We interviewed twenty easeouts: six males and fourteen females. Five of the males and thirteen of the females were African-Americans. Most of these females, both African-American and White, had attended Letsche Education Center because of behavior problems or pregnancy. Others described a process of quiet, yet forceful, persuasion to leave. As one male explicated it: 'This vice-principal told me when you are eighteen, you get kicked out. That's what he told me. If you are doing good and don't cause no problems, you can stay. I found out that wasn't true. But that was the word'. Feeling intimidated, he left. In retrospect, he felt duped.

Two types of students are likely to be eased out: troublemakers and pregnant students. Powell, Farrar and Cohen (1985, p. 140) generalize how school officials handle 'troublemakers':

Schools tend to deal with their troublemakers through an elaborate set of administrative procedures that move them from one status to another, gradually distancing them from the rest of the school. First come in-house suspensions, then a special self-contained in-school program. Sometimes the final step is a separate no-frills continuation school, and sometimes the reverse movement back into school is made if a student is showing signs of improvement. But the behavior of unruly students rarely gets better. (p. 140)

Continuation schools therefore appear to be the primary institution that ease out students.

The continuation school concept grew out of the Progressive reform fervor, as we noted in chapters 2 and 3, to serve 'young workers, aged 14 to 18 years, with four to eight hours per week of schooling' (Kelly, 1993, p. 37). In spite of good intentions, optimistic rhetoric, and hopeful promises, the continuation, or alternative, school idea over time became devalued.

Social reformers hailed continuation education as a humane, preventive response to these individuals' neglected needs. Yet by segregating rebels and failures from the mainstream high school, educators stigmatized them and the continuation program while easing their disciplinary load and scaring other students into relative conformity. (*ibid*, p. 66)

Kelly argues that 'comprehensive high schools send disengaging students to continuation (schools), thus masking their own true dropout and pushout rates' (*ibid*, p. xv). We found the same to be true in Pittsburgh. Nationwide, in 1990, some four million students attended public alternative schools, 'of which continuation schools comprise the single largest category'. More importantly, and the focus of Kelly's (*ibid*, pp. xv and xvi) research, only 10 per cent of the students in California's large continuation school network received diplomas. Kelly's data and analysis condemns the continuation school approach:

its position as stepchild of public education limits its effectiveness in engaging students most alienated by academic learning . . . Like the lower tracks within the comprehensive high school, it too readily becomes a dumping ground for rebels and the academically underprepared. Unfortunately, educational policy makers have been content to provide a 'second choice' to those not well served by the mainstream without necessarily demanding that it be a better chance. Indeed, continuation schools often offer diluted academic preparation and become stigmatized as second rate. (p. xvi)

Continuation schools therefore represent the 'wastebasket of compulsory education' (*ibid*, p. 68).

The Letsche Education Center serves as the alternative, or continuation, school in Pittsburgh. It maintains four programs: Special Twelve Program for Seniors; Educational Medical Program for Pregnant Teens; Make up Program, Grades 9–12; and Retrieval Program. Letsche also maintains a higher annual dropout rate, at 16 per cent, than any other building in the district.

Our informants attended Letsche for a variety of reasons. Dissatisfied with her high school, an African-American narrator opted for Letsche, but she was highly critical of her experience there: 'The atmosphere at Letsche was horrible. It was bad . . . The teaching was poor . . . They really didn't care. It didn't make a difference if you came to school or not . . . It wasn't even a challenge'. High school authorities sent an African-American male student to Letsche because of 'too many fights'. However, the district misplaced his cumulative records: 'When I went to Letsche, they said they sent my records (to his former high school)'. His former high school denied this. He returned to Letsche: 'They said they sent my records', but he returned again to his former high school which said 'they didn't have them'. He grew frustrated: 'There was a lot of confusion with my records and stuff. I didn't want to be bothered with it. So, I just got fed up and quit'.

At least two informants recognized Letsche's mission. An African-American female labeled it as 'an alternative school for bad kids. I just didn't fit in with them'. Another appeared ambivalent:

> Letsche is not no big high school . . . That's a last resort high school . . . It's really for bad kids. I only went there cause I didn't like the other high schools, cause they were too big . . . It (Letsche) was really small. The hours were shorter than other high schools. It was a very small building.

Neither felt any enthusiasm for the school.

Many pregnant students also attended Letsche, encouraged by social workers and others, as we have seen. Safety and flexibility clearly represented the reasons for sending pregnant women there. However, one White female criticized it, because she had to climb steps at Letsche, which she found difficult because of her pregnancy: 'Like three to five flights of them . . . But they were suggesting that they were going to put in some elevators. I don't know if they did'. The steps at Letsche just compounded her dilemma. She left school at age 16.

A Personal Perspective

One African-American narrator 'loved' her experiences at the 'Letsche alternative', seemingly disputing Kelly's findings.

> They let you work at your own pace. (Letsche) teachers were not always at the board, doing this and doing that. Coming out, breathing down your neck. I didn't like that. Down in Letsche, if they put some of those teachers in regular high schools, the high schools would be all right.

This student described, in detail, the pedagogical methods of one of her favorite teachers at Letsche:

> There was all kinds of tables, you know; there were four of us to a table. She would come to the table. She would help us do our work. She would laugh with us, and talk with us, you know. The class was an hour long. We had time to do our work and talk a little bit. If we needed something, she would always come. We get projects in the room. Her class was biology. And in biology in [my former high school], I didn't like it. But when I went to Letsche and did biology, I made an 'A'.

Her counselors helped her make the transition to Letsche, which she attended on five separate occasions.

> At Carrick, I liked my counselor . . . He got us into Letsche. Letsche helped us out you know. Everybody was already talking about how tired they was of school; how everybody was going to drop out. And once we got into Letsche, that brought me further. My counselor at Schenley . . . I liked him. He helped me go from Schenley 'back to Letsche. I think I know this about myself. I can only deal with Letsche.

She appeared to travel the classical path to this alternative program, with the support and assistance of counselors. Further, the teachers there seemed to be effective with her. Nevertheless, she still abandoned school, conforming the alternative-school pattern of easing students out.

Combinations

The disengagement process was even more complex and diverse for some students. One African-American female, on the surface, appeared to conform to the typical pushout pattern, but upon closer examination she recounted other experiences. She began by claiming to have been pushed out: 'I got kicked out in tenth (grade)'. However, as she continued she described a far more complex phenomenon. The pushing out process began, as it ended, in tenth grade. She had missed a great deal of school because she had a child at home: 'Half the time I couldn't find nobody to watch her'. Her daughter was 'too young' for a child care center. This set the stage for a confrontation with a teacher, which led to her suspension.

> I was failing the class and [the teacher] told me to do a report (and) he would pass me. I would just pass with a 'C'. So I said all right, and I did a fifteen-page report. It took me about two weeks, but I did it. And when I handed in my report, and I came back three days later for it, to see what I had got, he told me he lost it. And I just went off. I started throwing books and chairs at him.

She appealed to the school's dean: 'I told the dean. He said that he can't do nothing about it'. She believed the system and personnel were unjust. She also felt alienated from the personnel. Between ninth and tenth grades, she had attended three different high schools, causing her to observe: 'I didn't really know my teachers'. To add insult to injury, she did not receive word of the suspension until after she left school. 'That day, when I went home, it was the last period of the day, they had already called my mother and told my mother don't send me back. They sent a letter saying I was kicked out'.

That action sealed her fate. Having been held back in the second grade, she appeared vulnerable; the school made it easier for her. It 'really didn't make a difference because I had missed so many days. I would have failed anyway. I wasn't staying back in the other grade, 'cause I was seventeen in the tenth grade, and I wasn't gonna be eighteen in the tenth grade'. She gave up. The care of her 4-year-old daughter overwhelmed her; it interfered with her initial attendance as well as precluded her return to school. 'Without a child, I would do it', she mused. 'If I would have been able to bring my child to school with me instead of having to find a babysitter, it would have been better'. The combination of an overburdened single teen parent and a school's insensitive personnel, a seemingly arbitrary appeal system, and lack of child care facilities led to this student's leaving school. Her fading out culminated with a push.

Other combinations existed as well. An African-American student pointed out that his dropback decision stemmed from an easing out process. Because of his home problems, his high school counselor advised him to leave school and enroll in the Job Corps program.

Dropping Back

The typical dropback pattern proved to be incredibly simple, and often a result of serendipity. The process usually unfolded as this White female student recalled it: 'I quit school and I was looking through the *TV Guide* one day, and I seen an article about Job Corps. So, I called them, and a month later I was here. I have been here eighteen months now'.

Other manifestations existed as well, with school leavers repeating success, avoiding failure, and reflecting maturity. In the first case, a female student followed her brother's path into the Job Corps. He received training and found a job; she wanted to duplicate this positive experience. As she recalled: 'I didn't want to go to high school. I really didn't. I thought it (the Job Corps) was a good opportunity. My brother went here. He said it would be good for me. I'm going to college'. A second informant simply wanted to avoid being just another dropout. His mother prodded him to resume his schooling.

> I had dropped out and my mother had dropped out. She was telling me that she feels bad that she didn't go to the prom or got a class ring. I have been thinking about that a lot. I would see a lot of people. My mother would ride around and say, 'See him. He dropped out. Look at him. See him. He used to be this and he had this, and he had that. Now look at him'. That's what made me come up here (to the Job Corps).

He too planned to enroll in the college program. Finally, a dropback, who deeply regretted leaving school, planned to attend college. Reflecting maturity and toughness gained from experience, she summarized why she resumed her schooling: 'I know what I gotta do now. It's not easy out there in this world. You gotta work to get something. You can't depend on other people to give you something. You gotta make it on your own, 'cause there ain't gonna be nobody there all the time for you'.

For some dropbacks, returning to school was essential to fulfilling long-term plans and aspirations. They based this realization on short-term, limited work experiences. 'You need education out in this world', one African-American male proclaimed:

> If you don't have too much education in the work world, you are not going to get too far. If you get more education, you go further. That's what I see now. That's the only reason why I'm glad I came here (to the Job Corps). After I finish and get my GED, I'm going to a college program. I have to spend two semesters up there [at the Job Corps] and then transfer to West Virginia University. I'm going into the ROTC program down there. I plan to go into the service.

He often missed school to take care of his terminally ill mother, as reported earlier, and sought financial security and the promise of a stable future.

The Job Corps attracted most school leavers as a bridge to a job, but a few others liked the nurturing environment. As one African-American female described it:

> They help you. They help you. You say, well, I can't do this problem. They don't say, well, go to the next problem, and then go back to that problem. They explain it. They give you a book list that you can read. My mother was surprised at all the little certificates and awards I brought home from Job Corps.

These dropbacks also revealed how they viewed the role of education in American society; most entered the Pittsburgh Job Corps program in order to obtain a job. For them, education only served a utilitarian purpose. As one African-American female described it:

> Here at Job Corps it offers a lot of trades that you can get into. They weren't teaching me [in high school] how to get a job. I never knew what a resume was. I filled out a couple of applications before and I learned how to deal with being turned down. I know what I want to be now.

She planned to be trained for a maintenance position.

A Personal Perspective

A 21-year-old White female dropback, who had held five jobs since the age of 13, shared her insights into the dropping back experience.

School is too important to me now. Everybody thinks I'm so retarded over here (at the Job Corps). All the students have this fake macho image that, 'Hey, I'm tough. I can make it through anything', which is not true. You can't make it through anything if you don't have one foot in the door and ready to try.

She planned to attend college after her Job Corps training and she earned her GED: 'I'm sticking it out. I'm looking at long-term goals. I don't want to be a hamburger flipper at McDonalds. I want a job that I can be proud of, so I can say this is where I go to work everyday'. She clearly envied her former classmates who graduated from high school: 'I don't like it that they did something that I could do and didn't'.

Conclusions

The tragedy here is that the typical school leaver 'knows the reception that awaits him or her in the outside world, yet believes that it cannot be worse than remaining in school' (Kowalski and Congemi, 1974, p. 73). In gross terms, why school leavers left and why they returned can be stated in deceptively simple ways: they left because school was boring and antagonistic, they returned very largely for economic reasons. The interviews revealed subtle differences under each of these two headings that both enrich and complicate our findings.

The School as Boring and Antagonistic

Many of the interviewees described school and the associated academic activities as boring. Little struck them as interesting or important. A dull, uninspiring, alienating academic climate was often the salient factor causing students to leave school. Our informants detailed some of the very deficiencies outlined by major studies of American schools as Boyer (1983) and Goodlad (1984). These school leavers related classroom activities dominated by lecturing, and so-called seatwork and tests. In more instances than not, teachers were portrayed as mere dispensers of subject matter. They perceived teachers as uncaring. The teacher, a classroom authority, told students what was to be learned. Students, in turn, were to repeat that information in classroom recitation, or in passive seatwork assignments, and receive a grade for the unit in questions on a test or examination. Students did not feel that they participated in learning of subject. We asked school leavers to describe their good teachers. Those so identified — few and far between — were those who in one way or another engaged students, were open to questions, and worked with anyone who had trouble understanding the class lecture or the text. As some interviewers simply put it: 'they helped me'. What they left unsaid was that other teachers, usually a majority, did not. These students sat in class unengaged and did minimal classroom work, if any at all, and never had a sense of its meaning for them. This led to cutting classes, or 'fading out', prior to leaving school entirely. In sum, informants judged their classroom experience as dull or 'boring'. This sounds remarkably similar to the problems identified by early twentieth-century Progressive reformers, described in chapter 2, who played down poverty as a cause of leaving school and, instead, stressed boredom as the cause.

More than boring, many of those interviewed saw school as an antagonistic environment, reporting various hostile situations. Overall, students noted conflicts with other students and school personnel. They were often the victims of harassment. Both males and females reported incidents outside of school, some near, some far. Such incidents of harassment led to fights that would break out inside the school. This led to disciplinary action by school administrators, often after the intervention of school security personnel. 'School membership' appeared marginal at best and nonexistent at worst (Wehlage, 1989, pp. 2–3).

School leavers were equivocal when asked whether these disciplinary incidents were dealt with fairly. In some cases, they saw security personnel as fair intermediaries; some school administrators, notably school vice-principals with disciplinary responsibilities, also were said to be fair. But administrators making summary judgments were just as frequently perceived as arbitrary. There are residual questions which need to be seriously pondered and discussed: Are security personnel closer to the life situation of school leavers? On the other hand, are school administrators hostages to central administration to report that 'things are under control' in their building? How can teachers and other personnel establish contact with students who perceive the school climate as boring and antagonistic?

Why School Leavers Returned

The school leavers interviewed for this study participated in a reentry program. Between leaving school and enrolling in the Pittsburgh Job Corps, many, if not most, confronted severe economic problems. They realized that any employment without a high school diploma would be marginal at best. Many of them had held part-time jobs either after school, during summer vacation, or both. Such employment included minimum-wage jobs such as in fast food franchises and baby-sitting. In each case, the narrators recognized that they were unlikely to find steady employment at anything beyond the minimum wage. Consequently, they hoped that returning to a reentry program, obtaining the GED, and receiving job training would make them more employable. Their return to an alternative educational program was stimulated by economic concerns, not intellectual curiosity, which virtually none of our informants exhibited any evidence of (see Fisher, 1992, p. 21).

One may well ask whether schooling has any intellectual import for students like those we interviewed. They were not concerned with questions of the meaning of life or their role in the larger society — understandably so since life had already taught so many of them that theirs would be a marginal role. More immediate concerns dominated their lives. As we pointed out earlier in this study, they lacked the fundamental support systems that many of their more affluent peers enjoyed. Those in their immediate family occasionally lacked a sense of life goals. There seemed to be no reason to delay gratification, to study and achieve academic success, and benefit from it later economically or culturally. Very few turned to community neighborhood resources such as religious organizations and recreational programs; just as few took part in school activities.

As a result, many of the school leavers spent their formative years generally on their own, without contact with some 'significant other', drifting aimlessly through their development. In turn, once they left school, the question of supporting themselves became paramount. Their past job experiences had been at marginal levels, and the adults they observed, in and outside of the family, were too mired in unemployment or underemployment to serve as positive role models. So students returning to school through the Job Corps arrived with a newly realized goal of economic self-support.

In addition, the economic factor was further reinforced in the case of the teenage mother, now faced with the care of a child. In contrast to the limitations of school-sponsored child care, the Job Corps' child care program provided support for their own educational and vocational development unavailable elsewhere.

In sum, economic factors are paramount for school leavers who return to schools, quite different from those of nineteenth- and early twentieth-century experiences. Students once fled school to go to work; now they abandon work to resume their schooling. Whereas the workplace, not the classroom, once provided job training and supplied a viable living, profound changes in the political economy have fundamentally altered the relationships between families, schooling, and work. At the end of the twentieth century, work no longer represents an alternative; schooling serves as the bridge to a job.

Note

1   The data in these paragraphs are drawn from two district sources: 'Pittsburgh Public Schools Annual Report on Suspensions: School Year 1983/84', pp. 4, 10 and 11; and 'School Profiles, School Year 1992/93', Division of Student Information Management, January 1994. In the latter case, see the 'Secondary Schools' section, pp. 145–68.

Part III

*Conclusions, Analyses, and Policy
Implications*

Chapter 8

The Social Context

In terms of physical appearance and condition (lower-class) students disgust and depress the middle-class teacher. (Becker, 1951/52)

American educators indiscriminately used the terms *dropout, student elimination, withdrawal,* and *early school leavers* during the first half of the twentieth century. The descriptor *dropout* did not come to dominate educational literature until the late 1950s. High school graduation came to be regarded as essential only after the Second World War, an attitude peaking during the first half of the 1960s. School officials viewed dropouts as possessing flaws which prevented them from adjusting to school, and as unable to comprehend the educational and social costs of their leaving. Educators therefore saw dropouts themselves as the source of the problem (Dorn, 1993, p. 369). Their truncated schooling also contributed to youth unemployment. Missing from the extensive 1960s literature about dropouts was any major analysis of the American social structure, and the inequalities inherent in it.

The dropout issue decreased in importance until the early 1980s, when it reemerged as a significant social and educational problem. Although a larger proportion of students have graduated from high school since the sixties, American society has grown to expect even more of them to complete their schooling (*ibid*). Educators and critics have once again highlighted leaving school as a cause of unemployment among young people, and they have largely overlooked social theory as a means to explain the phenomenon of dropping out.

Five, often overlapping, themes emerged from our analysis of school leavers' perceptions of schooling during the late 1980s and early 1990s. First, social class factors have contributed to leaving before graduation. Poor and working-class students have abandoned school in far greater numbers and at higher rates than their more affluent classmates. Second, failures of pedagogy appear to be an important element influencing student disengagement from the process of their learning and from their appropriation of school knowledge. This too is related to social class. Third, student alienation, or disengagement, while certainly related to social class, can transcend it. Fourth, these alienated students demonstrated clear modes of resistance to schooling, that led to further alienation. Fifth, attitudes toward work permeates the historical and contemporary experiences of school leavers. This chapter addresses these five areas and, generates a theoretical framework to explain the experiences described by our informants.

Social Class

Social theorists, regardless of their political perspective, have long acknowledged the linkages between social class and schooling experiences. George Counts, a social reconstructionist, whose pioneer and now classic quantitative 1922 study, *The Selective Character of American Secondary Education*, which relied on a survey of urban high schools, found that 'educational opportunity' rested on the social class position of the student's parents. Class position was not only correlated to attendance and completion rates, with poor and working-class students maintaining lower graduation rates than their more affluent peers, but also correlated to curricular choices, with 'laboring class' students enrolling in vocational courses and wealthy ones taking college-preparatory classes. 'Thus secondary education', Counts (1922, pp. 142–3 and 148) concluded, 'remains largely a matter for family initiative and concern, and reflects the inequalities of family means and ambition'.

The continuities in these patterns over time are stunning. A.B. Hollingshead's ethnographic study of adolescents in a small, midwestern town in 1941 found clear social class distinctions in that community that were translated into the high school experience. First, social class expressed itself in curriculum choices. The local high school maintained three tracks: college preparatory, general and commercial. Affluent students gravitated toward the college-preparatory curriculum, while poor and working-class youths enrolled in general and commercial courses. Second, their academic preferences shaped the teachers' attitudes:

> Because the academic teachers believe that college preparatory students have more ability, are more interested, and do better work than those in the general course, they prefer to teach the former group . . . These teachers look upon the students in general courses as persons who have nothing better to do with

their time, are mediocre in ability, and lack motivation and interest. Students in the commercial courses are believed to be (even) lower in ability than those in the general course. (Hollingshead, 1975, p. 125)

Teachers therefore treated students differently according to academic track, which, of course, reflected social class biases. Third, students' course grades matched their social class position, with teachers awarding high status students with better grades and low status youths with poor grades. Fourth, school administrators meted out discipline along the same lines. They treated wealthy students in a rational and flexible way and reacted irrationally and rigidly to poor students. Finally, school leaving trends definitely followed social class lines. Hollingshead (1975, pp. 126 and 254) writes, 'The adolescents' own reasons for leaving school may be grouped under three headings: (1) economic need; (2) peer isolation and discrimination; and (3) mistreatment by teachers'.

This latter pattern continues to this day: social class inequalities directly affect school leaving patterns. 'National data confirms that social class is the most reliable predictor of dropping out for females and males' (Fine and Zane, 1989, p. 26; see also Elliott, Voss and Wendling, 1966). Social class even supersedes race as a cause of inequality. Holding socioeconomic status and student grades constant, Solomon (1989) observes, African-American 'students are less likely than their White and other minority-group peers to drop out of school' (p. 79). We therefore must look beyond the schools for solutions. Michael Apple (1989) maintains that as long as we remain fixated on the schools, we will remain unable to adequately address the school leaver problem.

Lasting answers will require a much more extensive restructuring of our social commitments. Further, they will need to be accompanied by the democratization of our accepted way of distributing and controlling jobs, benefits, education, and power. Until we take this larger economic and social context as seriously as it deserves, we shall simply be unable to adequately respond to the needs of youth in this country. (p. 206)

To exacerbate matters, poverty is increasing in the United States, and social class inequalities appear to be growing worse (*ibid*, p. 208). Will even more poor and working-class youths abandon school? How many will resume it later?

Our informants, with rare exceptions, failed to recognize such inequality in their lives. For example, they knew and acknowledged racism, but seldom pointed to systemic — either in school or in society — racism; they usually noted individual acts of racism on the part of their instructors and peers. Were they naive? Or, has the system worked, as Michelle Fine (1991) asserts, in 'anesthetizing' them? Fine's (1991) critique of social consciousness, her concern for 'racial, cultural, and class-based anesthetizing' of graduates as well as school leavers, is instructive. In a Gramscian sense, bourgeois hegemony works. Marxists, like Antonio Gramsci, use the concepts of false consciousness and hegemony to explain how the ruling class impedes class consciousness and progressive social change. When the subordinant class expresses the views and values of the dominant class as well as appear loyal to it, they exhibit false consciousness. When the dominant class establishes its way of thinking among the subordinate class, it establishes hegemony. For example, when workers who are exploited — overworked, underpaid, lacking benefits, and unprotected, — believe the system is fair, do nothing to change matters, and go so far as to espouse the values of the owners, they are victims of false consciousness and hegemony. It pervades all social institutions, especially education. 'Hegemony exists when one class controls the thinking of another class through such cultural forms as the media, the church, or the schools' (Feinberg and Soltis, 1985, p. 50). False consciousness and hegemony contribute significantly to social reproduction. In support of the views of 'radical educators', Giroux (1983) summarizes the broader processes of reproduction in schools.

> First, schools provided different classes and social groups with the knowledge and skills they needed to occupy thier respective places in a labor force stratified by class, race and gender. Second, schools were seen as reproductive in the cultural sense, functioning in part to distribute and legitimate forms of knowledge, values, language and modes of style that constitute the dominant culture and its interests. Third, schools are viewed as part of a state apparatus that produced and legitimated the economic and ideological imperatives that underlie the state's political power. (p. 258)

Pedagogy

Social class explanations provide the structural reality underlying the problem of abandoning school, but the class process becomes translated into the lives of students through daily pedagogy. The classroom interactions consistently described by our informants reflect more than just static pedagogy, as suggested earlier by Cuban (1984). Social class represents a key variable; teachers of low-income students treat them differently. Becker (1951/52) interviewed sixty Chicago teachers in the late 1940s, and found, in his now classic study, that 'social-class variations' existed in 'teacher-pupil relationships' (pp. 451–2). These instructors readily distinguished between 'lower-' and 'middle-class' students. The former, according to the teachers, came from 'slum areas' and usually attended 'slum schools'. These instructors commented on the teaching process, classroom management, and the students' 'moral acceptability', that is, their 'health and cleanliness, sex and aggression, and ambition and work', when they described their students. Lower-class children, for teachers, proved to be the most difficult to teach because they demonstrated little, if any, interest in their schooling, possessed low abilities, and lacked 'outside training' (*ibid*, p. 454). These same students presented teachers with control problems. Instructors resorted to overt and severe methods, like 'tongue lashings' and corporal punishment, in order to ensure order (*ibid*, p. 459). Finally, teachers felt little rapport with their lower-class students. 'In terms of physical appearance and condition, they disgust and depress the middle-class teacher' (*ibid*, p. 462). These instructors saw their job largely as custodial. However, teachers described middle-class students in more positive terms, albeit with some warts; the values and attitudes of teachers and students for the most part matched. Instructors also noted that middle-class parents tended to be intrusive, overseeing their children's schooling while their lower-class counterparts appeared to be apathetic, which partially explains these differences between teachers and their students. Other studies corroborate Becker's findings. Almost twenty years later, Rist (1970) found a differentiated schooling process in his ethnography of an urban elementary school. Social-class values guide pedagogy, as Becker concluded in 1951: 'All institutions have embedded in them some set of assumptions about the nature of the society and the individuals with whom they deal, and we must get at these assumptions, and their embodiment in actual social interaction, in order fully to understand these organizations' (p. 463).

Bowles and Gintis (1976, p. 132) also argue that schools treat students differently, depending on their social origins, but they transcend mere narrative description, developing a theoretical framework to explicate this process. They first point out factors which contribute to class-based differences in school socialization, and the major structural differences among schools. Schools serving working-class neighborhoods appear to be more regimented and place more emphasis on rules and behavioral control. Middle-class schools, in contrast, offer more open classrooms and educational experiences that 'favor greater student participation, less direct supervision, more student electives, and a value system that stresses internalized standards of control'. Such variations reflect the different expectations of teachers, administrators, and parents for children of different social backgrounds. On the one hand, working-class parents, from their own workplace interactions, know that submission to authority is an important value for success in the workplace; therefore, they often pressure the schools to inculcate this idea. On the other hand, middle-class parents, reflecting their work experiences, expect more open-ended, independent, and creative learning activities for their children (*ibid*, pp. 132–3).

Second, and most important, Bowles and Gintis (*ibid*, p. 12) rely on the 'correspondence principle' to highlight the similarities between the social relations of production and the social relations of schools. 'Specifically, the relationships of authority and control between administrators and teachers, teachers and students, students and students, and students and their work replicate the division of labor which dominates the workplace'. They argue that strong structural similarities can be seen in the organization of power and authority in the school and workplace; the student's lack of control of curriculum and the worker's lack of control of content of his/her job; the role of grades and other rewards in the school and the role of wages in the workplace as extrinsic motivational systems; and competition among students and the specialization of academic subjects and competition among workers and the fragmented nature of jobs. In sum, the social relations of the school reflect the capitalist mode of production. Schools help to tailor the self-concepts, aspirations, and social class identifications of individuals to the requirements of the social division of labor that exists in the workplace (*ibid*, p. 129).

Bourdieu (1977) goes beyond structural explanations; he instead analyzes cultural forms to find deeper social meanings. He defines 'cultural capital' as the general background, knowledge, disposition, and skills that one generation passes onto the next. Middle-class children bring a different background to the schooling experience than do working-class children. Family background provides linguistic and cultural advantages to upper-class students, giving them an educational edge over lower-class students. Middle-class students, therefore, appear to be more familiar with the dominant culture, which the educational system requires for academic success. Successful academic performance, as this argument goes, often results in the acquisition of superior jobs. Thus, the school, as a social institution, ensures social inequality through cultural reproduction (Bourdieu, 1977, pp. 496 and 507; Bourdieu and Passeron, 1977; Karabel and Halsey, 1977, p. 554; Swartz, 1977, p. 548).

Bernstein takes Bourdieu's theories a step further by analyzing both structures and practices and demonstrating their relationship. Bernstein meticulously traces how social class affects language use, an important component of cultural capital. He argues that social class shapes distinctive forms of speech patterns through family socialization. Working-class children exhibit 'restricted' linguistic codes while middle-class children express 'elaborated codes'. Bernstein sees linguistic codes not as the surface manifestation of language, such as vocabulary or dialect, but as the 'underlying regulative principles that govern the selection and combination and different syntactic and lexical constructions' (Atkinson, 1985, pp. 66, 68 and 74). Rooted in the social division of labor, linguistic codes are derived from the social relations within families. Working-class children, on the one hand, receive their primary socialization in homes where common circumstances, knowledge, and values produce speech patterns whose meanings remain implicit and dependent on their context, i.e., a restricted code. On the other hand, middle-class families utilize elaborated codes when expressing unique perspectives and experiences; they rely less on parochial and more on cosmopolitan symbols, less on the concrete and more on the abstract. Consequently, middle-class language operates as more linguistically explicit (Karabel and Halsey, 1977, p. 477). Since the social structure limits access to elaborated codes, and since the language used in schools is based on the symbolic order of elaborated codes, poor and working-class children operate at a distinct disadvantage. Thus, schools are simply not made for these children. Bennett and LeCompte (1990) summarize the research in this area:

> teachers of lower-class students employ custodial forms of behavior management; while much money and effort may be poured into special programs to enhance the achievement of these students, the programs often are remedial and simply repeat previous material; where computers are used they are used simply for an additional form of drill-and-practice. Placed in the lower tracks, less capable students get fewer hours of actual instruction and less rigorous coursework; 'pull-out' programs of special tutoring actually diminish the time spent in regular instruction. Teachers of the poor do not expect their students to do well, and, assuming that they will fail, interact

with them less, give them less encouragement, and worry less about the dropouts. (p. 172)

As our interview transcripts document, our informants experienced such a process, manifested through favoritism.

Alienation

School leavers in our study experienced a rather high level of school alienation which in turn contributed significantly to the abandonment process. Alienation theory explains why students fail to connect with the goals of the schools, develop a detachment from the schooling process, and eventually leave.

School attitude is of paramount importance in a society which relates social acceptance, individual worth, and social status to educational background and occupation. The problem of youth alienation is exacerbated when young people are not effectively being helped by society and its institutions to develop meaningful and satisfying educational and occupational roles, and, in effect, are being denied even the hope of becoming successful in either. Moreover, studies of alienation in public schools tend to view alienation as a static phenomenon that is best measured empirically. Such studies may supply evidence to support a large unifying theory, but sheds little light on alienation as a process which involves both teachers and students who seek to find meaning and purpose within an institutional context (Fensham, 1986, p. 5).

We believe our qualitative study avoids the limitations posed by quantification and allows the diverse voices of the school leavers to help shape its meaning and context. There is a richness and depth of feeling expressed by school leavers in this study that cannot be approximated by the collection of even large volumes of statistical data. The voices we have heard, recorded, and analyzed have provided us, we believe, with a clearer picture of why they have left school than have the statistical studies we have encountered in our review of the research. Those young people we interviewed universally articulated a sense of alienation from society, expressed a profound loss of interest in their studies, and became generally disillusioned. Most were bored with school and turned-off from learning. They pointed to uncreative teaching and an irrelevant curriculum, all of which detracted from the learning environment for students, teachers, and administrators.

It is clear that prior to several decades ago alienated students were less problematic than contemporaries, since most could opt-out of schooling for the workplace without much notice or thought about the consequences. In more recent times, entry level jobs for school leavers have shrunk considerably. In turn, a rapidly growing number of school leavers are becoming more involved in socially destructive behavior, from drug abuse to gang violence. The fall-out, therefore, from a high percentage of mostly lower class, mostly minority, students prematurely leaving school creates serious social problems. The major downturn in the employment opportunities for those prematurely leaving school differentially affects the students who most experience alienation from the schooling process (*ibid*, p. 20). The critical question, then becomes: How does alienation occur and how is it dealt with in the school community? Through the voices in our study, we have begun to address that question.

The classic descriptors of what constitutes alienation found most commonly in the literature are powerlessness, meaninglessness, normlessness, social isolation, and self-estrangement. Powerlessness is characterized as 'the expectancy of probability held by the individual that his (*sic*) own behavior cannot determine the occurrence of the outcomes or reinforcement he (*sic*) seeks' (Seeman, 1959, p. 784). Meaninglessness is defined as a 'low expectancy that satisfactory predictions about future outcomes of behavior can be made' (*ibid*, p. 786). Social norms fail to influence behavior when cynicism or normlessness sets in. Normlessness is defined as 'a high expectancy held by the individual that socially unapproved behaviors are required to achieve given goals' (*ibid*, p. 788). Although there are ways social isolation is manifested, the most common characteristic of the isolation is that they 'assign low reward value to goals or beliefs that are typically highly valued in the given society' (*ibid*, p. 789). The alienated cares little about commonly accepted rewards. Persons who are, in effect, alienated from themselves, and who have essentially lost contact with their basic needs, are considered self-estranged. The self-estranged feel that they have failed to realize their full human potential and that day-to-day actions no longer serve their basic needs. There is a further dimension of alienation referred to as cultural estrangement. Individuals remove themselves from mainstream cultural values, from the values of the community when they become aware of sharp differences between the common cultural environment and their vision of what culture could become or should be (Loken, 1973, pp. 220–3; Seeman, 1971, pp. 82–4).

On another level, Merton's (1938) study on *anomie* sheds light on what is meant by the term alienation. He described the alienated person as one who is 'in society but not of it' (*ibid*, pp. 673). The alienated student remains in school but has little or no interest in it, or has not thought of other alternatives, or fears change and clings to the familiar. This nearly always results in feelings that range from boredom to outright hostility toward the schooling experience. Merton (1938) also theorized that the alienated individual often becomes resigned to failure regarding the ability to reach personal goals. The alienated also frequently fail to adopt illegitimate means toward attaining their goals because of internalized prohibitions. This is because the alienated sub-culture has not fully renounced the dominant culture's goals for success. The individual solves this dilemma by eliminating the idea of both the goals and the means of achieving them. The alienated rejects the dominant culture's goals and values and substitutes new goals and values which are nearly antithetical to those of the dominant culture. Some students are alienated from a certain subject or set of subjects, such as in the math or the humanities curricula. Non-academically inclined students often believe they attend school merely to collect 'a piece of paper' that they believe will enable them to get a job. Other students attend school only because there is nothing else to do (Loken, 1973, p. 19).

Social conditions that foster alienation, of course, range far beyond the classroom. However, our research indicates that the school contributes significantly to the school leaver problem. Some students, particularly poor African-Americans and Latinos, in a certain sense are instructed to be alienated. They often learn by experience that there are different rewards for equal amounts of effort. In other words, the rewards for their effort are not commensurate with rewards given to other students for the same or similar effort. They feel unfairness, injustice, and downright discrimination. The alienated tend to form their own cliques which can eventually result in street gangs. Unfortunately, drug trafficking and even violent acts may also occur.

Disaffiliation, disassociation, and disengagement are terms used to describe and to modify the concept of alienation. However, these concepts can be as unclear and abstract as the concept of alienation itself. Very basically, as it applies to students, the term alienation, can be defined as an unsatisfactory relationship that exists between a group of students and the schools they attend (Loken, 1971, p. 21).

Viewed from a political perspective, alienation can have a revolutionary affect. Revel (1971) argues that society declines sharply because of an 'internal absenteeism', which often leads to alternative commitments. Those who adopt the new values of a different association are drawn mainly from those alienated from mainstream culture and the groups thus formed constitute a counter-culture (*ibid*, pp. 85–6). In a sense then, alienated students constitute part of the counter-culture of the school (Loken, 1971), as we shall see.

Eric Fromm (1955) regarded alienation as a person's loss of the sense of self in terms of his/her own particular powers and abilities. Alienated individuals are thus unable to identify with the larger social system to which they belong because they cannot exercise their own powers or express their own personality within the particular environment created by the system (Loken, 1971, p. 21; Fromm, 1955, p. 126). They tend to pull-back from situations which involve commitment which often indicates an attitude of scorn for society. Alienated students consciously reject the norms and values of schools. They are turned-off or lack sensitivity to the demands and incentives thrust upon them by mainstream culture, particularly the school. In turn, the alienated are attracted to new and different stimuli that create new sensations (Kenniston, 1965, p. 246).

Resistance

Alienation manifests itself in many ways. The alienated student may, on the one hand, quietly and passively withdraw from classes, peers, and finally school itself, as we have seen. On the other hand, alienation may result in resistance to schooling, which also leads to abandonment, though gradually through the pushout process. This latter sense of resistance permeated many of our interviews.

Willis (1977) terms this latter experience as 'opposition' (p. 22). Poor and working-class students, Willis found in his ethnography of twelve white male students in an English working-class school, formed a 'counter-school culture' to resist the formal structure of schooling (*ibid*, p. 23). 'The school is the zone of the formal. It has a clear structure: the school building, school rules, pedagogic practice, a staff hierarchy with powers ultimately sanctioned . . . by the state, the pomp and majesty of the law, and the representative arm of the state apparatus, the police' (*ibid*, p. 22). The counter-school culture, which directly opposes the formal school structure, is not always as visible or blatant.

Even though there are no public rules, physical structures, recognized hierarchies or institutionalized sanctions in the counter-school culture, it cannot run on air. It must have its own material base, its own infrastructure. This is, of course, the social group.

> The informal group is the basic unit of this culture, the funda-
> mental and elemental source of its resistance. (*ibid*, p. 23)

Members of this counter-school culture learn, through their informal
group ties, how to manipulate the formal system of schooling in a
variety of ways, demonstrating their rejection of imposed school authority and the entire schooling process.

Such opposition assumes many forms. Truancy represents the
assertion of 'informal student mobility'. Students, in this manner, actually control and arrange their own class schedules, if not all of their
schooldays.

> Truancy is only one relatively unimportant and crude variant
> of this principle of self-direction which ranges across vast chunks
> of the syllabus and covers many diverse activities: being free
> out of class, being in class and doing no work, being in the
> wrong class, roaming the corridors looking for excitement,
> being asleep in private. The core skill which articulates these
> possibilities is being able to get out of any given class: the
> preservation of personal mobility. (*ibid*, p. 27)

In this sense, students control their own time; their self-direction, of
course, assaults institutional and 'official notions of time' (*ibid*, p. 28).
'Having a laff' too serves as a form of resistance. Offering a release for
'boredom', it defeats 'fear' and overcomes 'hardship and problems'
(*ibid*, p. 29). This involves pranks and jokes played on teachers as well
as students; many of these acts represent 'antisocial' behavior. Fighting
serves as another distraction.

> There is positive joy in fighting, in causing fights through in-
> timidation, in talking about fighting and about the tactics of
> the whole fight situation. Many important cultural values are
> expressed through fighting. Masculine hubris, dramatic dis-
> play, the solidarity of the group, the importance of quick, clear
> and not over-moral thought, comes out time and time again.
> (*ibid*, p. 34)

These activities together operate as more than just amusement; they
'subvert' the formal school setting. This entire subversion process often assumes the form of the 'general rejection of school work', and
feigned 'stupidity' usually leads the list (*ibid*, pp. 30 and 32). A fine line
exists here, however. Members of the counter-school culture must

persuade their peers that they are indeed intelligent, or they will lose status within their group because they will be labelled as dumb, yet they must not academically succeed, or they too will experience rejection for their intelligence. Street smarts, therefore, supersedes book knowledge. The result is that failure in the formal setting is rewarded in the informal world of the adolescent.

This represents a cumulative process: alienation produces a counter-school culture which, in turn, encourages further disengagement and failure, which, of course, leads to more alienation. Willis (*ibid*, p. 3) indeed found 'that groups of working-class lads come to take a hand in their own damnation'. He describes a complex reproduction process in which the structure dooms the resistor to a subordinate structural position. Resistance ironically facilitates social and cultural reproduction.

Work

Students have come to expect that doing well in school translates into the common expectation that this will result in meaningful work. In spite of schooling serving as a bridge to work, the relationship between successful high school completion and finding a 'good' job has become less and less clear. Even those who find employment often receive a low paying wage. Custodians, cashiers, secretaries, office clerks, and sales clerks represent the five fastest growing occupations in the United States (Apple, 1989, p. 214). Nevertheless, the perception persists: students who do not necessarily like school are willing to endure that educational experience, and even make an effort to do well, simply because they believe it directly affects the quality of employment they will obtain. Teachers and administrators reinforce the importance of the relationship of schooling to work; thus, most students still believe in, or at least recognize, the utilitarian value of schooling. However, for those students alienated from the schooling experience and destined to leave school, the linkage between schooling and work has little meaning.

Bowles and Gintis (1976) maintain that workers under a capitalist economic system experience

> powerless(ness) because work is bureaucratically organized, ruled from the top, through lines of hierarchical authority, treating the worker as just another piece of machinery . . . to be directed and dominated. Meaningless because it is divided into numberless fragmented tasks, over only one of which the worker

> has some expertise, and whose contribution to the final product is minimal, impersonal, and standardized. Meaningless, equally because the worker who produces goods designed for profit rather than human needs realizes only too well how dubious is his or her contributions to social welfare. (p. 72)

Workers are usually isolated in their work, and the fragmented tasks they perform precedes solidarity and cooperation. Hierarchical authority pits workers on different levels against one another. The powerless, meaningless, and isolated position of the worker leads him or her to treat work merely as an instrument for attaining material security rather than an end itself, but work is so important to self-definition and self-concept that the individual's self-image crystallizes as a means to some ulterior end; hence, the worker's self-estrangement (*ibid*, pp. 129–32).

Central to Bowles's and Gintis's (1976) analysis is the correspondence principle, as we saw earlier. This Neo-Marxist argument is based on the assumption that capitalist societies reproduce through their structure and function the social relations of capitalist production. The correspondence theory places the production system of capitalism and class conflict at the center of an explanation of school/work relationship. It maintains that capitalists extract surplus value from wage labor in order to realize profit and to expand capital. To gain efficiency and control over the workplace, capitalists employ a division of labor, a fragmentation of the work place combined with hierarchical-technical control of the production process. The result is that workers are deskilled and lose control of both the processes and products of their own work. This is the basis of their becoming alienated human beings. The educational system helps to integrate youth into the economic system through a structural correspondence of the schools' social relations and the social relations of production. Social relations in schools not only helps to prepare students for the discipline of the workplace, but assists in developing personal demeanor, modes of self-presentation, self-image, and social class identifications which are crucial ingredients of the future workplace.

Schools play a crucial role in capitalist efforts to gain control over workers — to gain worker acquiescence to dominant hierarchical social relations. Schools aid in the reproduction of ideologies that reinforce the class differences that undergird capitalist production. They help to reproduce the division of labor through the introduction of segmented units of instruction, the use of curiculum packages that tends to deskill the work of the teacher and the students, and pedagogy that produces behaviors and attitudes of student conformity, i.e., work hard to gain approval and comply with the depersonalized rules of the school bureaucracy. Schools sort out potential workers, professionals, and managers in line with the social class structure through differentiated curricula and certificates. In short, schools help reproduce the social relations of production — to develop the kinds of skills and attitudes that produce a docile, exploitable labor force (Wirth, 1983, pp. 165–6; Bowles and Gintis, 1976).

More specifically, the social relations of schooling — between administrator and teachers, teachers and students, students and students, and students and their work — replicate the hierarchical division of labor. Hierarchical relations are reflected vertically from administrators to teachers to students. Alienation is reflected in the students' lack of control over their education, the curriculum content, as well as the grades and other external rewards, rather than the student's integration into the learning process or knowledge outcomes. Fragmentation is reflected from competition among students through meritocratic ranking and evaluation (Bowles and Gintis, 1976, p. 131).

Levin (1982), on the other hand, argues that the correspondence principle emphasized by Neo-Marxists is an exaggeration that does not mesh with the complexities of American society. He maintains that tension exists between two social forces in capitalist societies, which helps to explain the mix of correspondence and the lack of correspondence between work and school. The state is viewed as an arena of struggle between the needs of capital and the needs of labor. The legitimacy of the state is dependent on providing popular reforms, while, at the same time, creating policies to accommodate the needs for private capital accumulation.

As a part of the state, schools reflect the same contradictions, i.e., the need to meet the demands for popular, egalitarian reforms, yet providing support for capital accumulation. The schools are thus in opposition to themselves by being organized to serve the needs of two powerful groups with conflicting goals. Schools attempt to meet the needs of both the democratic and egalitarian aspects of schooling and the authoritarian and hierarchy structures needed for reproducing labor for capitalists. Schools come under the control of neither capital nor labor. This explains how schools correspond to both the workplace hierarchy and opportunities for social mobility through schooling (*ibid*, pp. 19–20; Wirth, 1983, pp. 166–7). Levin maintains that the movement towards greater worker participation and collective decision making could have significant implications for schools (see also Kerchner and Koppich, 1993).

Because of their close ties, we believe there needs to be a direct relationship between the reform of the workplace and the reform of schooling. The dialectic between work reform and school reform has the potential for a more positive response from at-risk-students. When the schooling experience has little relationship to meaningful work, at-risk-students are less likely to stay the course until graduation. Providing linkages between workplace and schooling experiences that make sense to at-risk-students can have an impact, stemming the tide of student school leaving.

Conclusions

Social theory, as we have presented it here, raises fundamental questions about the relationships between school and society, and cautions against a business-as-usual mentality regarding public schooling. Accordingly, we ask several questions: First, should school reform be rethought in terms of the utility of school structure and function for those alienated? Our research found that early school leavers found little, if any, support in the schools as they are presently constituted. School leavers perceived that school personnel rarely dealt with their problems, that is, they saw no place where their problems could be presented and treated to their satisfaction. Second, is the simple identity of school reform with improvement in academic achievement an adequate measure? Informants faced a multitude of problems ranging from peer harassment and boring and irrelevant instruction to the absence of adequate support from their homes, communities, and school personnel. Under these circumstances, how could they have achieved? Third, can the schools facilitate the transition from school to work? The literature on this issue appears to be extensive, and we have done little more than scratch the surface here; yet, it has become apparent to us that this is an area which deserves greater attention. Many narrators in this study, when expressing an evaluation of their curricular instruction, saw their academic experiences as unconnected to their present realities and their futures. Are these students acting out their alienation from schooling because it is training them, as Bowles and Gintis (1976) suggest, for subservient work roles? Or, is Levin (1982) correct when he argues that the correspondence between school and work appears to be ambiguous? As long as students are not given a sense that their formal education will lead to better life chances, then their alienation from schooling will not be mitigated. Can school personnel clearly provide potential school leavers with an understanding of how their academic studies can lead to worthwhile work? Finally, and related to all of the above questions, the longstanding issue of equal educational opportunity needs to be revisited. Given the realities of our social structure, do poor students truly experience equal educational opportunity? Can we challenge and recast our unstated assumptions about the relationship between public education and equal, or should we say 'unequal', opportunities? These questions articulate fundamental issues we need to confront in the context of the early school leaver.

The results of this study, as well as other school leaver analyses, cry for alternative and/or augmented schooling strategies. The public schools, for the most part, appear inflexible and static while students' needs and society's demands have profoundly changed. However, Kelly (1993; see also Fine, 1991, p. 176) cautions how the process of sorting continues within the context of offering educational alternatives:

> offering options without major reforms both inside school (for example, improving the material conditions of teaching, rethinking standardized curricula as well as traditional teaching and disciplinary strategies) and outside the educational system (for example, providing job programs, more and better child care, funded access to contraceptives and abortion services) can simply mean replicating the sorting process pioneered in traditional educational settings. (p. 217)

Simplistic solutions, therefore, will only change the form of schooling, but not the substance and structure. Solutions must be comprehensive, acknowledging the complexities of the school leaving process. They must transcend the limits of the existing paradigm of public schooling, and seek creative and flexible approaches, overlooking nothing. All of this must begin with an atmosphere of caring and sensitivity.

Chapter 9

Caring

Nobody gave a damn. (Pittsburgh school leaver)

When we began this research, we did not have a clear sense of what it might identify. We thought that early school leavers might have found the school environment hostile or uninteresting. Weren't there better things to do? Perhaps, it seemed lucrative to deal in drugs, according to one stereotype of school leaver as deviant. Or, given the statistics on teenage pregnancy, other things occupied them. We had picked up these notions from the popular press and other studies. However, beyond such vague ideas and some other reasoned hunches, we were searching for clues.

As our interviews unfolded, we began to see that one factor stood out. There were, of course, immediate problems like pregnancy, truancy, harassment, boredom and poor achievement. Still, as we looked at interview transcripts cataloging such problems, it was not until we asked another question that things started to fall in place. What was missing? Was there something lacking in the experience of these school leavers? Then what was common in the interviews became apparent. We had seen again the validity of the remark attributed to Einstein: 'Imagination is more important than knowledge'.

What appeared eloquent in earlier and later interview transcripts was the importance of caring, or more to the point, we observed a lack thereof. As one student declared: 'Nobody gave a damn'. Here it should be noted that lack of caring was not attributed to the school environment and school personnel alone. Apparently 'nobody gave a damn' about these young people outside the school — at home, in the neighborhood — as well as in school. Therefore, we devote this chapter to an analysis of 'caring' as a basic ingredient in mitigating the incidence of school leaving.

This is not to single out or indict teachers, other professionals, or any single group. Still, our findings report that early school leavers did not have a high regard for most teachers. School administrators and other professionals, such as counselors were more often than not perceived as 'shadow' figures. Given the heavy load borne by counselors and the large responsibilities of school administrators, such findings come as no surprise. Nevertheless, their indifference represented a void in the life of the early school leavers. Whether that void can be filled is a significant question for teachers and school boards and central administrators as they manage fiscal affairs and personnel policies.

We will analyze four interrelated aspects of caring: schooling and a caring climate, fairness as instrumental to caring, a philosophical analysis of what it means to care, and implications for teachers and pre-professionals.

Schooling and a Climate of Caring

We initially now focus on the element of caring in school practice, the subject of a growing body of literature. For example, Newmann (1992) states:

> There is more to life than academic achievement. Academic success must not, therefore, be the sole criterion for school membership. Students' moral worth and dignity must be affirmed through other avenues as well, such as nonacademic contact between staff and students . . . In short . . . the separate features we identify (purpose, fairness, support, success) must be integrated within a more general climate of caring. (p. 23)

As such, 'students are cared for as persons who represent multiple aspects of humanity, not simply as units to be processed through the official agenda of the school' (*ibid*).

Newmann (1992) suggests conclusions similar to our findings.

> Regardless of the level of extrinsic rewards, students may invest in or withdraw from learning, depending on how interesting they find the material. Interest refers to the fact, that some topics and activities are considered more stimulating, fascinating, or enjoyable to work on than others. What will be interesting, probably depends not simply on the subjects or topics, but largely on the way the topics are approached by the teacher, the student's prior experience with similar material. (p. 25)

What is important is the student's 'sense of ownership'.

Engagement with and internalization of knowledge depends to a large degree on the opportunities students have to 'own' the work. Rather than toiling under predetermined routines to master skills and knowledge dictated arbitrarily by school authorities, students need some influence over the conception, execution and evaluation of the work itself. (*ibid*, p. 25)

What can schools do? Here the analysis displays some clear positive features of constructive schooling and some problems to be encountered in policy and practice. Foremost is a connection to the real world. As Newmann notices, at least four qualities of adult work in the real world that are often missing in school work: value beyond instruction, clear feedback, collaboration, and flexible use of time (*ibid*, p. 26).

It is apparent from our research that early school leavers were disengaged from school work because they did not see its relevance to their lives. Was this their fault? In these few instances where the intervention of a teacher made some dent in the level of students' interest, for example, when the teacher was characterized as a 'big brother', it was clear that an effort was made to relate subject matter to the learner's frame of reference. What disturbs us is that such caring was so seldom evident. Newmann's four qualities of instructional concern defined caring for educational professionals at whatever level they serve. How can one learn without pointed feedback? Isn't that what is entailed in teacher–student collaboration? And doesn't that require flexible use of the teacher's and student's time?

Fairness as Instrumental to Caring

But there are pitfalls and problems to be dealt with. So we move from caring as providing a nurturing climate to another key element: fairness (on justice as fairness, refer to Rawls, *A Theory of Justice*, 1971). Newmann (1992, p. 32) describes a typical incident. A boy who has been repeatedly picked on is finally forced to fight his tormenter in self-defense. Both boys are apprehended and receive the same punishment — 'four days suspension from school and three days of school detention'. One of the victim's teachers agrees that student suspension seemed too extreme, but rationalizes it: 'If you punish everybody you don't have to decide who's innocent and who's guilty. You just throw everybody into detention. That simplifies the problem for the staff' (quoted in Newmann, 1992, p. 32). Newmann's commentary is enlightening:

The problem should be understood from both angles (school responsibility and student/family responsibility). But there is a long standing tradition in American education of blaming students for their own failure and marginalization. This perspective has led both educators and policy makers to ignore the positive role schools can play in creating an educational environment that builds student membership and offers students authentic work that produces high levels of engagement and achievement. (*ibid*, p. 33, see also Newmann, 1991; Newmann and Wehlage, 1993)

In short, school professionals — teachers, counselors, social workers, and administrators — need to ask whether their judgments and actions are fair to affected students. This requires caring about and sensitivity to the life situation of the students involved in such incidents.

What It Means to Care

In an extended analysis, Nel Noddings (1984) distinguishes between the one-caring and the one cared-for. She considers whether caring should necessarily be visible or have some behavioral manifestation. If caring is to take some overt form, then it might be expected that rules for caring can be identified. In fact, the search for and the stating of such rules has been the traditional approach in ethical investigations.

However, Noddings (*ibid*, p. 9) argues: 'I think emphasis on actualization of the other may lead us to pass too rapidly over the description of what goes on in the one-caring'. Instead of looking at caring from the outside, her approach is from the inside, that is, as a complex phenomenon caring is as much for what involves the one-caring as the one cared-for. She notes, 'if we can understand how complex and intricate, indeed how subjective, caring is, we shall be better equipped to meet the conflicts and pains it sometimes induces' (*ibid*, p. 12). In turn, this means: 'What we do depends not upon rules, or at least not wholly on rules — not upon a prior determination of what is fair or equitable — but on a constellation of conditions that is viewed through both the eyes of the one-caring and the eyes of the one cared-for' (*ibid*, p. 13).

Fundamental to caring from the inside is apprehending the reality of the other, what Noddings (*ibid*, p. 16) refers to as *engrossment*:

Apprehending the other's reality, feeling what he feels as nearly as possible is the essential part of caring. For if I *take on the*

> *other's reality* as possibility and begin to feel its reality, I feel, also, that I must act accordingly; that is, I am impelled to act as though in my own behalf, but in behalf of the other. (emphasis added)

For a teacher, such engrossment does not come easily. It requires more than surface observation of a student's situation. It may also lead to potential conflict and/or a difficult decision point. Does one care for student A at the expense of student B? Can one care for both A and B? What about C, etc.? Or does one take the line of least resistance (or least difficulty) as in the case of suspension cited by Newmann above?

But what if the cared-for individual does not appropriately respond? Suppose one cares for a student, tries to enter into his or her frame of reference, but the response is negative? Or, the student may ignore the caring overture. Between the one-caring and the one cared-for there needs to be reciprocity. And one must take the risk of rejection if caring is to occur at all. The overture needs to be genuine. Moreover, each of us at times needs space or even solitude. That should be sensed and honored by the one-caring. So, a critical element in caring is timing. Thus, understanding the other's frame of reference, empathy, risk taking, and timing are important.

Caring by definition involves feelings. Accordingly, there are no rules, no recipe. What is required is willingness, concern and empathy. In other words, a caring attitude does not come cheap.

Another critical fact is the difference in socioeconomic status (SES). If there is a difference in SES between teacher and student, communication can be disrupted and, in turn, a caring overture may be misinterpreted. To wit, in a study of a remediation program in Buffalo, New York, Bram Hamovitch (1993) reports just such a problem. From interview transcripts, he reports that the

> narrative reveals the communication between staff, students and parents is far from complete. Even though the participants do not express themselves using the concept of 'social class', it appears that the parent . . . senses a middle class arrogance and control, while the staff member interprets a lower-class incompetence and uncaring attitude toward the child. The lower-class mother does not feel that she is able to communicate on an equal footing with the middle-class staff member, while the latter feels that the former is being deliberately evasive in . . . communications with her. The dynamics of the relationship, including the crossing of class barriers, certainly adds a

dimension of unpredictability and potential misunderstanding that is likely inherent in many such relationships. (p. 8)

Class size represents another impediment for a teacher attempting to find ways of relating to students. As class size is increased, the teacher's ability to reach all students is stretched, probably with negative consequences. Thus, such impediments to teachers' caring need to be recognized and removed if the situation of early school leavers is to be mitigated.

The notion of reciprocity has already been stressed. Then what about the one cared-for? Noddings (1984) recognizes that 'the cared-for must turn freely toward his (*sic*) own projects, pursue them vigorously and share his accounts of them spontaneously' (p. 75). A great deal is involved here. We need to carefully unpack the reciprocity between the one-caring and the one cared-for. Consider therefore a problematic situation where the one cared-for may not respond in the way the one-caring may wish or anticipate. Noddings quotes Urie Bronfenbrenner: 'In order to develop, a child needs the enduring, irrational involvement of one or more adults in care and joint activity with the child' (*ibid*, p. 61). The word 'irrational' raises questions that Bronfenbrenner answers by stating: 'Somebody has got to be crazy about that kid'. Noddings goes on:

> The child about whom no one is 'crazy' presents a special problem for teachers. Obviously, the teacher cannot be 'crazy about' every child . . . but the teacher can try to provide an environment in which affection and support are enhanced, in which children not passionately loved will at least receive attention and, perhaps, learn to respond to and encourage those who genuinely address them. (*ibid*, p. 61)

One can well question whether being 'crazy' about someone is irrational, or even whether that is an apt synonym for caring. However, Bronfenbrenner's hyperbole does get our attention and it says, especially to professionals in education, that caring is central to their task. More to the point of this study, Noddings's comment about the importance of a supporting school environment is borne out in our interviews.

In sum, the caring and cared-for relation is central to addressing the school leaving problem. From a strategic standpoint, the issue is how to view the schools' role in this process. Noddings's approach is critical:

> Possibly no insight of John Dewey's was greater than that which reveals the vital importance of building educational strategy on the purposes of the child. The principle of the leading out of experience does not imply letting the child learn what he pleases; it suggests that, inescapably, the child will learn what he pleases. That means that the educator must arrange the effective world so that the child will be challenged to master significant tasks in significant situations. The initial judgment of significance is the teacher's task. (*ibid*, p. 63)

Dewey's view has frequently been misconstrued in the movement for 'child-centered' schooling. As if analogous to free love, so-called free schooling has been thought to make curriculum and instruction subservient to the whims of individual children. Such a notion, however, misses the point of Dewey's philosophy. His view of social control emphasizes the need for rules governing human activity. The rules are not arbitrarily imposed but are part and parcel of the activity as in a game (Dewey, 1963, p. 52f). Hence the teacher explores and discovers these activities commensurate with students' capacities for learning in a setting where the rules for proceeding are apparent and agreed to by all parties, or, at least, a majority of those involved. If that is accomplished, then a caring relationship has been established. What one would strive to achieve is what Noddings (1984, p. 78) describes: 'A caring relation requires the engrossment and motivational displacement of the one-caring, and it requires the recognition and spontaneous response of the cared-for'.

Still another analysis that emphasizes a caring educational climate is found in H. Jerome Freidberg's revision of Carl Rogers' classic *Freedom to Learn* (1994). Instead of focusing on problems that inhibit learning and cause students to leave school, Freidberg illustrates those factors that promote a love of school. On the basis of anecdotal reports of students, Freidberg reports: 'I found common threads that were consistent from one school to the next and from one school to another. Looking from the students' perspectives, I discovered links that bound them and their facilitators into remarkable learning communities' (Rogers and Freidberg, 1994, p. 5). Freiberg notes eight factors associated with such learning communities: students want to be trusted and respected; want to be part of a family; want teachers to be helpers; want opportunities to be responsible; want freedom, not license; want a place where people care; want teachers who help them succeed, not fail; want to have choices (*ibid*, pp. 5–7).

In Freidberg's research, students made such comments as: 'They

talk to us, which is, you know, a lot different than talking down to us';
'This is really our home, I mean, I am here (i.e., at school) more than
at home'; 'On a personal basis (the teachers) go to each individual and
ask how you are doing. Some people are going to be at a different
(academic) level than another individual'; 'I think our freedom is more
freedom of expression than just being wild and having no self-control.
It's like we have a purpose, and so our freedom is freedom to express
ourselves'; 'Most of the teachers here really care about me. They help
not just with the subjects they teach, but with other subjects and personal things. It is different than other schools where they tell you to
get your mind off anything that is not their subject'; 'The teachers care
about your grades . . . (and) help you out a lot. . . . You can go up to
them, they listen to you, they support you'; on the question of choices
in learning, 'This school has internships . . . We can leave to work . . .
Every Thursday, I work at the museum half a day . . . I found it helpful' (*ibid*, pp. 5–7).

All of these student comments emphasize how much students need
a caring learning environment, identified in this research as a major
element lacking in the experience of early school leavers. What this
suggests is a need for a profound cultural change among teachers and
in schools. Instead of viewing such findings as adding another responsibility for teachers who are already burdened, they should be seen as
offering an opportunity to mitigate the problems of early school leavers.

Implications for Teachers and Teachers-to-be

All of this may seem quite daunting to teachers: they already face
burdensome demands. In addition to providing a high and sensitive
level of instruction for developing minds and personalities, they are
occupied by a series of time–consuming duties, such as monitoring
lunchrooms, supervising playgrounds at recess time, filing reports,
and holding student and parent conferences. All of this, in addition to
planning lessons, devising relevant and meaningful assignments, and
teaching every day, can easily divert their attention and bring about
that fatigue labelled as 'burn out'. Ironically, demanding that teachers
add *caring* to the list could mean this necessary ingredient to quality
instruction would either be lost in the shuffle, that is, ignored, or
mechanically dealt with as another 'duty'. But instead of adding another layer to teachers' responsibilities, caring for students should per-
vade all activities and responsibilities.

This is easier said than done. Do teacher education programs prepare those anticipating teaching careers for such caring? Are pre-service mentors sensitive to this need? Do in-service and staff development programs highlight the importance of caring for students? Can caring be seen as attention to the poor achievers, the unruly, and undisciplined, as well as to the 'good' students? These questions must be asked *and* answered.

The difficult process of learning, identifying problems, devising solutions, and struggling to make sense of what we encounter in life experience, requires the help of other people: parents, friends, teachers, counselors, and other mentors. Without others who care, a young person will suffer. That recognition on the part of education professionals is basic to caring.

One practical value of Noddings' analysis is its emphasis on the reciprocal relation between the one-caring and the one cared-for. One must be prepared, however, for being rejected or ignored. It may seem that the caring gesture has failed. Instead of talking about failure, however, we would rather talk about readiness. It could well be that the one-caring is ready for a caring relationship, while the one cared-for is not. No doubt many teachers have experienced such encounters, but that does not represent failure or an excuse to give up. Noddings' earlier point is well taken. Providing a caring and supporting environment is a constant responsibility.

Teachers must try to know enough about the psychology of the students they are teaching to recognize the ways in which they seek acceptance and the kind of social and emotional problems they face. Especially situations of diversity in socioeconomic status and concomitant differences in cultural outlook, such as Hamovitch describes, require a knowledge base and sensitivity to enable one, as Noddings (1984) puts it, to 'take on the other's reality' and 'learn to respond to and encourage those who genuinely address them' (p. 61). In practical terms this suggests that attention needs to be given to the need for caring throughout the teacher's preparation. Pre-service teacher education needs to begin developing sensitivities in this regard. In-service and staff development must continue to deepen such sensitivities in the light of the actual, day-to-day experience of practicing teachers.

A caveat is appropriate here. Our reliance on an educational philosopher such as Nel Noddings needs to be put in perspective. She analyzes caring relationships in depth. We recognize that teachers cannot and in many cases need not get to such depth with their students. What they can do is be supportive, which can take the form of simple civilities; being pleasant, approachable, willing to talk, sensitive to mood shifts, noticing students' appearance. A colleague put it well: 'One doesn't have to be a psychologist to fill this role — just a common-sense person with antennae out to capture the "unsaid" in students' lives.'

What is being suggested here should not be misunderstood as a recommendation for teachers, counselors, and administrators to be permissive, warm and fuzzy, or soft and sentimental. It is more accurate to describe the caring attitude and behavior that is needed as a kind of 'tough love'. Many of the school leavers we interviewed have 'street smarts'. They learned survival skills — at best, counter-productive and at worst, antisocial in the long term — which only perpetuates their present alienation. The kind of sympathetic adult they need is one who firmly and positively says 'you can be better' and 'we are here to help you do just that'.

In our experience, that is what the Job Corps personnel are attempting to do. The staff members we worked with, including instructors, counselors, custodians, secretaries, and administrators, viewed themselves as counselors. Such a climate is needed in all schools. The enormous task of public education is to provide that kind of humanistic support. It means that there is a need to find ways to mitigate the charge of the least privileged that 'nobody gave a damn'.

Conclusion

In a time of educational reform, when emphasis is placed on international competition in school achievement, when discussion centers on national testing and strategies that include student performance, such as outcomes-based education (OBE), when some would have the government provide tuition vouchers for independent schools as alternatives to public schools, our research tells us that these issues and proposed solutions to education's problems remain irrelevant for early school leavers. Or, as Kenneth Boulding once put it, when commenting on the media theories of Marshall McLuhan: 'He has hit a very large nail not quite on the head'.

School reform has assumed many manifestations in recent decades. Great emphasis has been placed on student achievement. Critics have bemoaned the putative failure of United States students on standardized tests. Alarm has been voiced about an increase in drug and alcohol abuse among school students. Teen pregnancy is seen as another symptom of degenerating moral values. Racism and sexism continue to be denounced as they are found in schools. Violence in school, especially the growing incidence of weapons, haunts us as a people.

The litany of complaints and problems is extensive and often baffling to those responsible for the conduct of the education of the public. One of the more recent responses has been to deemphasize the central role of tax-supported public schools. It has been argued that if schools were forced to be more 'competitive' with each other, then the cream, would rise to the top. Why not institute mechanisms for funding students at non-public schools? Others have argued that the public schools need to be restructured. The bureaucratic management model appropriated from industry in the first half of this century, which views the school as analogous to a production unit like a factory, needs to be reformed (Callahan, 1962; Kerchner and Koppich, 1993). Overall, public rhetoric in Congress and in statehouses across the land calls for school improvement in proposals such as 'Goals 2000: Educate America'.

Whether such proposals can help students-at-risk remains an open question. Some of the rhetoric surrounding these initiatives is notably indifferent to the early school leaver. How, for example, so-called school choice could mitigate school leaving is unclear. One is tempted to argue cynically that proposals for enabling parents to choose their children's schools in a 'free market' avoids the problem of students who do not complete their public education by simply ignoring them.

But we as a society ignore this problem only at our peril. Our talks with early school leavers revealed just how little they could depend on the kinds of support systems that are so necessary to young people approaching adulthood, that are so important in preparing them for productive and satisfying lives. Here we have focused on one dimension — caring — and not so modestly suggest that education professionals must profoundly change their basic attitudes toward the students they serve. History has shown us that when society does not care for the least of us, we allow embarrassing problems to fester — the early school leaver, the homeless, disadvantaged minorities — only to come back to haunt us. So, shall we prize a caring attitude? Or shall we blame the victims? Our schools and society confront a critical choice.

The caring that concerns us here, which Noddings and Newmann talk about, albeit in different ways, is very simply stated by Cornel West in *Race Matters* (1993a). He shows how nihilism can be described as 'the murky waters of despair and dread that now flood the streets of Black America' (p. 12). He clarifies: '*Nihilism is to be understood here not as a philosophic doctrine that there are no rational grounds for legitimate standards or authority; it is, far more, the lived experience of coping with a life of horrifying meaninglessness, hopelessness, and (most important) lovelessness*' (p. 14). West (1993b) further noted in an interview: 'The kind of nihilism that I'm talking about in *Race Matters* can best be responded to not with grand vision and sharp analysis — even though they are important — but with very concrete expressions of love and care, to convince persons that they are actually worth something' (p. 1).

Our general conclusion is that a major theme for school reform is how to create and sustain true learning communities. Newmann emphasizes that a caring/learning community must be a total environment — both cognitive and affective. Noddings emphasizes the interaction and interdependence of the one-caring and the one cared-for. Rogers and Freidberg identify the positive aspects of educational development that, at least in part, can help to define a learning community. West (1993b) points to a fundamental human need: 'Concrete expressions of love and care, to convince persons that they are actually worth something' (p. 2). Such are the challenges for those who would address the problematic situation of the early school leaver.

Chapter 10

Policy Recommendations

If I could be a teacher for one day, my students would love me. They would love me. (Pittsburgh school leaver)

Our goal in this study was to reconstruct how schooling is experienced by school leavers, to gain insight into their culture and how they came to abandon and later resume school. Focusing on the city of Pittsburgh's public schools, we interviewed 100 dropbacks over an eight-year period. We analyzed this data by placing it against historical experience, applying social theory, examining philosophical frameworks and comparing our findings with those of other local and national studies to discern how they fit into a broader context.

Attitudes toward school attendance and completion have been mixed in American history. Some poor and working-class parents, because of concern for the family's survival or because they rarely saw any inherent value in school learning after a certain age, reluctantly, and seldom, sent their children to school during the nineteenth century. By the early twentieth century, compulsory attendance and social pressures caused parents to keep their children in school through adolescence. Progressive school reformers also created special programs to entice and hold students who might otherwise have left to work. This did not ensure completion, however. Some instructors and administrators begrudgingly taught students whom they did not necessarily want or welcome. Many students responded by leaving school. For parents and children for whom schooling offered no immediate, tangible rewards, it possessed little utilitarian value. These children left at the first opportunity.

Yet, compulsory education is not the cause of this purported crisis. Rather, the problem lies in the rigidity with which schools define their role. The school structure, of course, has attempted to keep students in school by hiring counselors and social workers to address the ever growing complexities of an expanded and captive student body in an increasingly threatened society. Nevertheless, even though schools have committed financial and personnel resources, their responses generally represent a superficial approach, treating the symptoms but not the causes of the school leaver problem. To put it simply, everything else has remained static. School officials still expect students to complete their public education within a twelve-year period. Those who do not are viewed as atypical or abnormal. The hours and length of the school day has remained the same for decades as well as the nine-month academic year. An urban, technologically advanced society continues to maintain a nineteenth-century agricultural schedule as we enter the twenty-first century. The whole concept and structure of schooling needs to be revisited and critically analyzed.

American society, for all of its vaunted ingenuity and efficiency, has not implemented the recommendations of earlier, informed studies of school leavers. For example, Beck and Muia made the following appeal in their 1980 dropout survey:

> The key to lessening today's high dropout rate . . . is early identification of dropouts and appropriate treatment of these children when they are young . . . If guidance counselors and teachers can recognize potential dropouts, show them approval and love, provide educational experiences for them . . . ensure some degree of success in their work, and help them overcome their academic handicaps, they will save these children a great deal of unhappiness, frustration, and hostility. (p. 72)

Yet little has changed since 1980, because our narrators, through the late 1980s and early 1990s, consistently described an environment that lacked these nurturing characteristics, an environment that condemned them to educational and, ultimately, social failure. As one former student summarized his school experiences: 'I'm gonna be honest; just going was the worst thing, getting up in the morning, staying there all day, boring too'. These narrators initially responded to an unfeeling and often unfriendly school environment by leaving it. Only by enrolling in the Job Corps, a costly remedy to society and a long slow road for the average student, can they return to school.

This society has for too long, taken uninformed and inefficient approaches to the school leaver dilemma. Humanistic reasons aside, allowing students to leave school before they have finished is just plain impractical: 'Letting a youngster quit school is the most costly and wasteful extravagance allowed by American taxpayers' (Kowalski and Cangemi, 1974, p. 72). Merely reacting to the problem rather than addressing its causes head-on is expensive and counterproductive. 'Because of unemployment, dropouts are more likely than graduates to require and receive public assistance' (Beck and Muia, 1980, p. 69; see also Markey, 1988, p. 41). As Rumberger (1991) adds, 'the economic costs alone amount to more than $200,000 for individuals over a life-time and more than $200 billion for a one-year cohort of dropouts' (p. 64). Even relatively successful reentry programs, like the Job Corps, are comparable in cost to private university tuition (Kowalski and Cangemi, 1974, p. 72). Money does not appear to be scarce, but we need to understand how it is used. This society each year commits enormous resources to the construction of prisons. The overwhelming majority of prisoners, who are disproportionately minority and young, left school before graduation. The irony here is that prisons are rapidly becoming 'the' high schools for African-American youths.[1]

Early intervention, therefore, is the key to preventing school leaving, and a caring and nurturing school environment, as discussed in the previous chapter, represents the most important ingredient in keeping students in school. Such investments in the reduction of school leaving could reap many rewards. The benefits to the individual student are obvious. Society profits as well. One study Rumberger (1991) cites reveals that 'expenditures on at-risk students' will result in 'higher earnings' (p. 85). Simply calculated on the 'increased tax revenues generated from these earnings, taxpayers will receive almost $2 for every dollar invested in dropout prevention'. Another study cited by Rumberger, which emphasized the 'social benefits of reduced crime, welfare, and training costs associated with dropouts, estimated that the benefits of dropout prevention would exceed the costs by a ratio of 9:1'. In lieu of this, reentry programs must be expanded. This chapter treats student graduation and reentry, dovetailing our findings from the Pittsburgh study with other state and local trends and programs.

Predictable Findings

Some of our findings could have been predictable. Other studies in the growing body of literature on early school leavers report similar findings. Still they are worth mentioning here, for they catalog, in part, symptoms of students who are at risk of leaving school.

Absenteeism

Many of our narrators told us without prompting about their progressive absenteeism from school. Their decision to drop out was not sudden. Many began by cutting classes. They usually explained their behavior as the result of boredom, and failure to see any relevance between what was taught and their lives. This applied to all fields — no one academic subject stands out. Most of the students claimed to dislike a subject because of dislike for a teacher. The key factor was whether the teacher made a connection between the subject matter and themselves. As time passed, their absenteeism grew from cutting an occasional class to cutting whole days. Consequently, they fell behind their peers and had less and less incentive to catch up. In other cases, violence contributed to absenteeism. Some students simply feared school and began to avoid it. When the bulk of our narrators looked back over their school experience, they often painted a picture of 'fading out'.

Low Achievement

By their own admission, many of the school leavers we interviewed (with a few surprising exceptions) did not do well in one or more subjects. At the same time, they saw little relevance or value in what they were taught. We do not make a wholesale indictment of teachers, yet we sense that some teachers did not take pains or were unable to show the relevance of what they were teaching and apply it to the life-world of the students. There appeared to be little if any dialogue about how the meaning of the material being studied could make a difference for the students, could affect their experience. Students were left to accept and remember whether or not they understood what had been taught. In many cases this led to progressively lower achievement and less and less contact with what transpired in their classes.

Alienation

The majority of students we interviewed were part of that overlooked underclass which inhabits large sections of urban America. With their plight largely ignored, with no one in their immediate experience to look up to, it is no wonder that they feel alienated from mainstream life. The reasons behind student alienation are varied. Clearly some students were alienated from school because of their poor achievement. Others were detached from their peers. They were picked on and threatened by other students and saw no way to defend themselves, nor anyone else to help defend them. In turn, they felt no reason to remain in school — just the reverse. Without some support, some help, some incentive to continue schooling, they simply left. School personnel rarely, if ever, encouraged them to stay; all too often they facilitated the students' departure.

School Suspension

Some of our narrators had been suspended from school, sometimes for short periods, others for longer stretches of time. We have no way of evaluating the fairness of these suspensions, yet it is worth noting that in a few cases the students themselves did not question their suspension per se, but attributed it to their general disenchantment with their school experience.

How effective is the practice of suspension? What do students learn from it? While one can concede the necessity of removing a troublemaker from a classroom, simply for the sake of the other students, what influence may it have on the suspended student? Does the troublemaker receive adequate counseling for understanding the misconduct in question? Are any steps taken to help the problem student or to change behavior? Typically not. Consequently, one suspension leads to another.

Teenage Pregnancy

Not surprisingly many female students left school because they became pregnant; however, the extent and scale of teenage pregnancy did surprise us. While the Pittsburgh Public School District has a program in a special school for such cases, the Ed Med program at Letsche Education Center was perceived as a 'dumping ground', and thus was not highly regarded. Subsequently, after delivering their babies, these young mothers were frustrated in attempts to secure child care if they were to return and complete high school. These teenage mothers could find no child care openings in the school district's limited child care program, and opted instead for the Job Corps program which does provide child care.

Recommendations Based on the Findings from this Study

Tinkering with the technical side of schooling, reducing class sizes, even caring about students, and expanding counseling personnel and facilities, while all of these are useful and humane attempts to address the problem, they will not resolve the school leaving problem alone. These tactics would produce only the 'illusion of change'. Fundamental structural problems exist because, 'at the same time, the deep structure of schools — firmly rooted beliefs about the nature of knowledge and learning, the purposes of education, and students' intellectual capacities, as well as the politically charged intersections of race, social class, gender, schooling, and life chances — go unquestioned and unchanged' (Kelly, 1993, p. xiii). These structural inequalities have existed throughout American history. The school leaving phenomenon, as we have seen, often reflects these inequities. The ultimate resolution to the school leaver dilemma lies in a fundamental restructuring of American society. But this is an unreachable goal in any immediate sense. This leaves us with 'systemic solutions' and 'reentry' policies, reflecting a functionalist approach, and, as Jonathan Kozol demonstrates in *Savage Inequalities* (1991), money makes a difference. Those politicians and policy makers who have claimed that a lack of money is not a problem in the nation's schools have simply chosen to disregard those inner-city schools with overcrowded classrooms, under-maintained buildings, and inadequate staff.[2]

Systemic Solutions

Systemic solutions rest on the fundamental notion that school leaving, as Rumberger (1991) terms it, 'is more of a social than an educational problem' (p. 84). Many students bring a variety of problems and crises to school with them that often leads to their abandonment of schooling. Rumberger continues:

> There are three major sources of influence on dropping out and other behaviors of young people—families, schools, and communities. If one views these sources as additive — that is, that each plays a significant role in influencing the attitude, behavior, and academic performance of young people — then each must play a role in addressing the dropout problem. (*ibid*)

Rumberger, therefore, proposes a comprehensive approach. Programs that hope to reduce school leaving must include parental involvement in fundamental ways, as well as 'strengthening the role of community organizations and the business community' (*ibid*). He concludes that 'some observers argue that families and schools cannot and should not shoulder the burden alone'. Wehlage *et al* (1989) summarize the cumulative effect of these experiences:

> If one comes from a low socioeconomic background, which may signify various forms of family stress and personal difficulties, and if one is consistently discouraged by the school because of signals about academic inadequacies and failures, and if one perceives little interest or caring on the part of teachers, and if one sees the institution's discipline system as ineffective and unfair, but one has serious encounters with that discipline system, then it is not unreasonable to expect that one will become alienated and lose one's commitment to the goals of graduating from high school and pursuing more education. (p. 37)

School Climate

Chapter 9 argued that early school leavers typically need somé kind of safety net to support them. Although those we interviewed did not complain about their family circumstances, they were left to fend for themselves at a fairly young age and rarely reported having looked up to a significant adult role model.

Who can fill this void in their lives? Cannot some 'significant others' care for them? One might expect teachers to play such a role. The sad fact reflected in this research, however, is that, for the young people we interviewed, teachers generally did not. For teachers to play such a role, it would require a change in behavior and attitudes. Such changes necessarily mean concomitant changes in the teacher's workplace, the classroom. School board members and district and building administrators need to recognize that it is their responsibility to provide a school climate that is safe and secure for both teachers and students.

Many changes can occur at the district level. Rumberger (1991, pp. 80–2) points to 'programmatic solutions', which include basic skills training, support services, and job-oriented assistance. More important, school districts must realize the linkage between professional devotion and alleviating student alienation. Firestone and Rosenblum (1988, pp. 285 and 286) analyze student and teacher commitment in particular, drawing on ethnographic data from ten urban comprehensive high schools in Baltimore, Newark, Philadelphia, Pittsburgh, and Washington, D.C. They identify commitment with 'attachment': 'Where such an attachment exists, the committed person is expected to believe strongly in the system's goals and values, comply with orders and expectations voluntarily, be willing to exert considerable effort beyond minimal expectations for the good of the system, and strongly desire to remain part of that system' (*ibid*, p. 286). This is a crucial relationship, because each group creates its own subculture, which appears 'mutually dependent yet in conflict' (*ibid*, p. 288). In the absence of commitment, and concomitant attachment between teacher and student, alienation is the result. Firestone and Rosenblum elaborate this process for teachers:

> Teachers' rewards typically come from knowing that students learn what is taught to them. When these rewards decline because students lack the commitment (as well as the requisite skills) to respond appropriately in class, teacher commitment is bound to suffer. In fact, teachers often complain about the problem of teaching apathetic, passive students and student ability is one of the most powerful factors determining teachers' sense of efficacy. (*ibid*, p. 289)

Teachers feel a strong sense of obligation when they interact professionally and productively with colleagues, receive administrative support, maintain high standards for students, face clear administrative expectations, and have autonomy and discretion over 'day-to-day decisions'. Firestone and Rosenblum (*ibid*, pp. 289 and 291–2) then turn to students, and found, through limited interviews, the same pattern we discovered, that is, students point to 'respect' and 'instruction' when defining a 'good' as opposed to 'bad' teacher. Student affiliation therefore depends on the teacher's professionalism. 'What they want is respect, the knowledge that they are being treated with decency and fairness by the adults'.

One cannot underestimate the importance of a positive school environment in keeping at-risk students in school. Our narrators saw school as threatening and unstable. Teachers spent most of their energy just keeping order. Students told us that many of their classmates brought weapons to school. A decline in school security and threats to safety are growing phenomena in urban schools.

Class Size and Access to Counselors

The students we interviewed frequently represented their high school as large and noisy. Although class size per se was not an object of complaint, the students did convey a sense that they were somehow lost in the shuffle. The school, for them, was not an attractive or friendly place. More often than not they felt a distance, even disengagement, from school personnel.

Debates about ideal class size are common in discussions about education. Especially in collective bargaining negotiations there is a tendency to allow increases in class size in order to offset personnel costs — fewer teachers per students, lower personnel cost. We conclude, however, that classes must be made smaller if authorities are serious about reducing the school leaver rate. Teachers with large classes do not have time for individualized attention to at-risk students. School district budgets reveal the systemic priorities. If the reduction in school leaver rates is a priority, then class size demands close attention with an eye to reducing the ratio of students to teacher and students to counselors. In this regard, the Pittsburgh schools report relatively low class size. While there is an average of eighteen students in secondary school classrooms, that figure is unrealistically low — including much smaller classes for exceptional students, for example, special needs classrooms and classes for scholars and/or high-achieving students.

At the same time, virtually none of the students we interviewed recalled any significant contact with school counselors. As reported earlier, counselors were 'shadow' figures. A few saw them as helpers, or someone to go to with a personal problem. This is apparently less the fault of the counselors than of the counseling system. It is not surprising that our narrators seldom if ever took their personal problems, academic or otherwise, to a counselor who may be responsible for about 350 students.

Flexibility

To correct the problem of students abandoning school, education professionals must devise more creative pedagogical methods and look for alternative scheduling patterns. School staff must be encouraged to refine and refresh their skills, and institutions must become more flexible. One of our informants pleaded, 'give us work experience, to try to go out and know what we are going to have to do when we get out there. Make the classes more fun or something. It's boring just sitting there doing papers all day'. Another narrator, totally without prompting, offered a positive solution through a moving description of her ideal school:

If I could, I would one day be a principal, and my school wouldn't be like . . . I have to have a school that I know every-body in there has an open mind, ready to learn. That gives a teacher a better outlook, ready to teach somebody who is go-ing to sit there and learn and listen. I even get teachers, who have a teacher aide, to sit there and help some of the students. I really would!

These students yearned for a humanistic school environment.

Other studies corroborate what we found in Pittsburgh. Gadwa and Griggs (1985, pp. 11–12) report the results of a Learning Style Inventory, conducted among 103 school leavers in the Edmonds School District, in Lynnwood, Washington. This inventory incorporated 'twenty-three variables relating to the affective, physiological, and cognitive domains'. School leavers responded to a '104-item self-report questionnaire that identifies learning preferences with regard to immediate environmental conditions and the students' emotional, so-ciological, and physical needs'. School leavers appeared to be moti-vated to learn, but poor pedagogy had discouraged them. Easily bored with 'highly structured learning requirements', these students preferred 'tactile, kinesthetic, and auditory perceptions as strong modalities in the learning process'. They wanted to be active learners, favoring 'va-riety in the learning environment' (*ibid*, p. 14).

Other potential and effective prevention models exist. Conrath (1984), a former teacher, generalizes about the 'defeated learners' he encountered at a Portland, Oregon, high school: 'School was very threatening. Classes were too large. Other students seemed hostile. Teachers appeared to be remote. Counselors were too busy. The Vice Principal was constantly on the lookout to "bust them"' (p. 36). Conrath, with the support of the Building Principal, implemented an alternative approach at this school. Potential school leavers attended a daily course titled 'Personal Growth'. This small class, with no more than twelve students, encompassed work on academic skills and gen-eral problem-solving, i.e., 'coping', strategies, introduced personal prob-lem sessions, and devoted each Friday to public congratulations on student successes. This intervention model also included weekly con-ferences with each student and emphasized student goal setting. Conrath and his assistant supplemented this with a comprehensive network, covering all aspects of students' lives:

When not with groups we met individually with students; worked with counselors, administrators, and the attendance

> office; talked with and encouraged parents; informed teachers
> of goal progress that related to their classrooms; got informa-
> tion from teachers that helped us monitor goals and assist stu-
> dents; and met together to develop materials, discuss students,
> and plan. (*ibid*, pp. 36–7)

This intensive experiment reaped positive results in a short time, with
school leaving rates decreasing sharply.

   Any solution to the school leaver problem must be flexible, because students abandon schooling for a variety of reasons. This complex reality demands sensitive policies.

> As states and school districts continue to devise and implement
> plans to reduce high school dropout rates, it is essential that
> policy makers take a more critical look at the diverse factors
> that may lead people to become at risk of not graduating. They
> must devise programs that respond in appropriate ways to that
> diversity.

Many students, for a variety of reasons, such as immaturity, problems
in development, and personal crises, need access to public schooling
for more than twelve years, although some may need less and many
may need to interrupt their schooling at times. 'The twelve-year-limit
on elementary and high school attendance — reinforced by state funding formulas — is unrealistic' (Bennett and LeCompte, 1990, p. 251).
This directly applies to young women who abandon their schooling
because of pregnancy. Farrell (1990) explicates this experience:

> Current social problems have made teenage parenthood a prob-
> lem. Poverty and lack óf vocational training make it impossible
> for most teenagers to support any children they may have. That
> there is little free education for people over twenty, and that
> there is an inadequate amount of day care for children, makes
> it difficult for mothers ever to be able to be breadwinners.
> (p. 89)

The hours and length of the typical school day must be reconsidered
as well. A more flexible school schedule could enhance attendance.
This is not a reward for adolescents who do not like to get up early in
the morning, but could give all students, not just potential school
leavers, options for work, meeting family obligations, and providing
child care.

Meeting such problems requires money. Reducing class-size especially mainstream classrooms — leads to higher personnel costs. So does reducing the counselor–student ratio. Can the social problem represented by a rising incidence of school abandonment move people to provide more caring personnel to relate to at-risk students? Are taxpayers willing to provide funds adequate for the needs of these students?

Child Care

A significant number of the females we interviewed left school because of pregnancy. Many would have returned to school if there had been in-school child care. The Pittsburgh school district does provide child care for teenage mothers, but the spaces are limited to about 250 slots, inadequate for the need. Many of the female students we interviewed chose to attend the GED program at the Pittsburgh Job Corps Center precisely because it provided child care. One can only speculate that there are others who do not find options like the Job Corps and do not return to school because they cannot find child care. There are no signs that rates of teenage pregnancy and the need for child care will diminish. Accordingly, a major reason for females leaving school before completion is inadequate child care.

Debates continue to rage over how to address the issue of pregnant teens, particularly whether or not the solution lies in a mainstream approach. Those opposed to permitting the pregnant student to remain in school are concerned about threats to the safety of the expectant mother and the unborn child, about subjecting her to hostile or insensitive comments, and giving the impression that the school condones adolescent sexuality and pregnancy. Proponents of mainstreaming, on the other hand, argue that pregnant teenagers should not be forced to make a traumatic transition and are less anxious about the problem of stigmatization; they contend that remaining in the mainstream allows these young women to maintain established friendships and continued access to curricular choices (Kelly, 1993, p. 221).

Young mothers' school experiences following childbirth presents them with yet another major obstacle. Our female narrators, as described earlier, needed support systems that would allow them to continue or resume their schooling. The most critical need they suggested was daycare. Without it, the rate of leaving school among pregnant adolescents will continue to escalate. 'Most teenage mothers chose to keep their babies, and most find few incentives and little assistance to remain in school'. The programs which exist appear to be effective: 'Programs which establish infant-care on public school campuses have been notably helpful in assuring a continued school attendance and as well as better child care among young mothers' (Bennett and LeCompte, 1990, p. 251).

Providing child care by the public schools can be costly. It adds significantly to a district's personnel budget. Moreover, trying to adjust to the needs of the student-mother could necessitate a more flexible schedule for students involved. In any case, without the provision of child care the correlation between teenage pregnancy and leaving school before completion will remain high. Child care by other family members is not the solution. Either other family members, such as a mother, aunt, or grandmother, are employed, or they are incapable or unwilling to assume a daytime child care responsibility. Therefore, in the case of these early school leavers, if the school itself does not provide daycare, they have little choice but to stay home.

Volunteer Mentors

Some potential early school leavers will fall between the cracks, even given improved instructional and administrative systems. We interviewed a sizable number of students who could not navigate the crosscurrents of the urban high school. Individual attention was lacking in their school, in their community and frequently in the home. A partial solution to the problem can be found in programs that provide surrogate support systems.

Los Angeles relies on the *Each One-Reach One* mentoring program that matches community volunteers with at-risk students.[3] Typically the students participating in the program have experienced and demonstrated poor attendance, low self-esteem and motivation, and poor grades. One of the program's coordinators, describes its character: 'The mentoring relationship is one of the few opportunities for a child to have 100 per cent of an adult's attention for an hour, for the sheer purpose of relaxing, expressing themselves and sharing companionship' (Collister, 1994, p. 1). One of the mentor's experiences underlines the importance of the program for the students, despite their initial skepticism:

> Angel's attitude toward me for the first six weeks was: 'He's going to get tired of doing this and quit coming' . . . Then when I returned after his one-month summer break, Angel

could not believe it. He had this confounded look on his face. After that he accepted me. You can't fool a child on emotional matters. (*ibid*)

The project manager, in turn, points out: 'Corporate involvement . . . is critical to the success of the program. Students at risk of dropping out are in desperate need of exposure to individuals who are "making it" in the workplace and leading productive lives' (*ibid*). As a result of his mentor's efforts, this student gained a new lease on life. His teacher reports:

These are experiences we can't provide in the classroom. . . . When Angel returns to class, he sits on the floor and shares . . . with classmates. This gives him an opportunity for verbal expression. He becomes the center of attention in a good way, not because he's gotten in trouble. And he gains his classmates' respect because he's doing something special. I've seen a real change in Angel. Now he's much more responsible and expressive. (*ibid*)

Many school districts have volunteer programs of some sort. Many have also developed links with corporate sponsors. We recommend that such programs be expanded so as to target at-risk students and provide them with a consistent one-on-one mentor relationship. The cost of such a program is modest. What is required is administrative support services to maintain an office, or a segment of an existing office, to recruit, coordinate, and monitor the match of volunteer mentors with at-risk students.

Reentry Programs

Our interviews demonstrate that, federal reentry programs like Job Corps work. Job Corps grew from Part A of Title I of the 1964 Economic Opportunity Act, a key element of President Lyndon Johnson's War on Poverty (Cremin, 1988, p. 316; Spring, 1976, p. 204). This legislation 'placed a new emphasis on youth' as a means of breaking the 'cycle of poverty', a popular notion at that time (Spring, 1976, p. 206). Administered by the Department of Labor, it attacked youth unemployment by removing at-risk adolescents, between 16 and 21 years of age from their home environments and placing them at residential settings, 'where they received lodging, clothing, meals, and health care, supplemented with training in basic education and vocational skills' (Cremin, 1988, p. 319; Spring, 1976, p. 207). This represented a combined public-private venture. The Department of Agriculture and the Department of the Interior oversaw the centers located in rural areas 'and engaged in conservation work'. Corporations, like International Business Machines and the Westinghouse Corporation, operated the urban centers (Cremin, 1988, p. 319). This 'expensive', yet attractive, program enrolled 400,000 young men and women alone between 1966 and 1972 (*ibid*; Spring, 1976, p. 298).

The Job Corps, as alluded to above, stresses academic and vocational skills. All centers give placement tests to determine achievement levels and maintains a standardized curriculum and small classes, with a maximum of fifteen students. They follow a rigidly sequenced and self-paced academic program, only allowing corps members to leave a course after mastering the subject matter. Teaching involves close monitoring, rapid feedback, and constant reinforcement; competition with classmates appears to be deemphasized. Participants usually earn a GED.

Job Corps concomitantly orients students about work conditions and potential earnings in various occupations, and allows them to experiment with tools and equipment before they select their vocational training. The Pittsburgh Center offers training Job hunting skills in Corps members also work six weeks with local employers to gain work experience. Some students pursue pre-apprenticeships. Participants who complete the program report receiving higher salaries as well as noneconomic benefits as a result of their Job Corps training (Weidman and Friedmann, 1984, p. 35).

Taking a comprehensive approach, the Job Corps assumes responsibility for the health and welfare of members. Counseling services, health care and child care overarch all of the academic and training activities. Trainees receive 'support and encouragement from the staff' (*ibid*). They learn self-government and perform housekeeping chores in the residential facilities. They can use weekend transit passes to visit friends or family. In addition to a modest monthly clothing stipend for themselves and dependents, participants receive a modest remuneration, which Job Corps invests in a savings plan; this is awarded after the program is completed. According to Job Corps, no student may remain in the program longer than two years. The Job Corps placement staff and state employment agencies assist job searches. Employers receive an enticement through the Targeted Jobs Tax Credit Program. Approximately 5 per cent of Job Corps trainees nationwide continue their education through college.

However, the Job Corps faces an ironic existence. In spite of its laudable record, its funding is challenged with each federal budget cycle. As one report has noted, the Job Corps

is widely hailed for its effectiveness and cost-efficiency, a Federal program that actually works. But despite rare consensus from government officials, elected leaders, economists and social scientists across the ideological spectrum, the program, which serves 62,000 young people a year, gets scant attention and struggles to hold its own in the annual budget minuet on Capitol Hill.

This shortsightedness reflects a general strategy taken regarding all social problems.

This is unfortunate, because the program appears to be successful. At the Pittsburgh center, according to the most recent audit, 62 per cent of the corps members obtain the GED in less than one year while 61 per cent of those who completed the entire Job Corps program have secured employment. Trainees are counseled by virtually every Job Corps employee. As we alluded to in chapter 9, students do not necessarily have to schedule formal appointments, but can be advised in all phases of the program by instructors, counselors, custodians, secretaries, and administrators. Such counseling, or caring, as we have recommended, follows each student. This 'tough love' does not tolerate all student behaviors; rather, it demands the best from each student (US Department of Labor, 1990, pp. 15 and 149).[4]

Caring Revisited

In the previous chapter, the importance of caring, among teachers and other personnel as well as the development of a caring and supportive school climate, was analyzed. Here we frankly admit that there is no magic formula for achieving such an intended aim, but that is not to suggest that the quest for caring by individuals and schools be abandoned. Common sense indicates that caring for others necessitates caring for self. So self-awareness development needs to be an important element in any effort aimed at strengthening and enriching a caring attitude in school personnel. Just as important, negative attitudes related to race, gender, age and social class need to be identified and analyzed in terms of their negative consequences for self and others.

Can such a caring attitude be developed and enhanced? If teacher preparation programs are not geared to the development of a caring approach to instruction, then this is an obvious starting point. If administrative certification programs are similarly not geared to the development of a caring school climate, then this represents another obvious area. Also, district in-service experiences serve as yet another arena in which self-consciously developed caring attitudes and outlooks may occur. Typically an emphasis on the development of a caring attitude for education professionals is not a feature of preparation and in-service programs. Its importance, however, by now should have become apparent. So we recommend that caring be a subject of serious consideration for in-service workshops and in training programs for teachers and administrators.

A caveat is also in order. A caring attitude and its attendant behavior, as we have pointed out, need not be soft and fuzzy. In the interviews we have conducted with early school leavers, what seems to be needed is not behavior typified merely by warmth and unqualified acceptance. Instead, as much as attitudes can be described, what is needed is clear and consistent support which focuses on student learning.

Conclusions

A profound reformation of the school's humanistic, academic, and structural environment is needed. This does not represent an original idea. Bryk and Thum (1989, p. 75 and p. 377), in their study of 'school effects on dropping out', point to a similar 'constellation' of 'structural and normative features', such as

> smaller high schools where there are substantial opportunities for informal adult-student interactions, where teachers are committed to and interested in working with students, and where students are pursuing similar courses of academic study within an environment that is safe and orderly. These are institutions whose structural and functioning coalesce around a sense of shared purpose. The result is a coherent school life that is apparently able to engage both students and teachers alike. Such strongly chartered schools offer one constructive approach to student alienation from schooling.

Our narrators, based on their school leaving experiences, express similar goals and described similar ideals, educational environments. What has struck us as profound is that much school leaver literature remains replete with such recommendations, but policy makers have tended to ignore them.

Rather than harp on the negative aspects of schooling, that is, what causes students to leave, we prefer to close our analysis of this social and educational problem by reviewing successful efforts to create learning communities. Wehlage *et al* (1989) studied fourteen such programs, and generalized from them. These schools, in a variety of urban settings, focused on two vital components to serve 'at-risk' students. First, teacher behavior played a crucial role, focusing on 'reciprocity', that is, creating a climate of mutual teacher-student respect. This research pointed to four other beliefs:

> teachers accept personal *accountability* for student success; they believe in practicing an *extended teacher role*; they accept the need to be *persistent* with students who are not ideal pupils; they *express a sense of optimism* that all students can learn if one builds upon their strengths rather than their weaknesses. (*ibid*, p. 135)

Second, school structure, involving size, autonomy, flexibility, and control, shapes students' attitudes towards schooling. Wehlage *et al*, define a small school setting as a program servicing no more than 500 students.

> To promote both school membership and academic engagement, it is essential that students have frequent contact with adults; in particular, it is through one-on-one relations that care, support, and personalized teaching are possible, and adults can come to understand students' problems and points of view. (*ibid*, pp. 143–4)

Autonomy refers to freedom of teachers to act in the best interests of at-risk students. In large measure, this means freedom from a lock-step curriculum, that is, teachers need the flexibility necessary to introduce new material and to adapt or even supplant the old. 'Without exception, educators cited autonomy as significant in their ability to construct programs that respond to students' (*ibid*, p. 144). Flexibility encompasses virtually all aspects of the educational experience. 'Different strategies are needed and these require flexibility in the development of curriculum, in the scheduling of student time, and in determining the actual site of educational activities as well as in the evaluation of students and the awarding of credits' (*ibid*, p. 144). Finally, teachers must be in control over the conditions of their work. In those fourteen programs, they participated in the selection of students and colleagues, coupled with 'site-based management with decision-making close to the scene of action' (*ibid*, p. 147). Such a nurturing approach builds 'a community of learners' who want to remain in school (Rogers and Freidberg, 1994, pp. 5–7).

However, the most important point is this: creating such learning communities not only benefits potential school leavers, by promoting their graduation and thereby easier access to more rewarding work or further education, and society, by reducing the costs of this social and educational problem, but such schools enrich all children.

Notes

1   This statement is based on comments made by Ron Cohen on a panel, 'Symposium on the History of Urban Education: Perspectives from Recent Works', annual meeting of the History of Education Society, 3–6 November 1994, Chapel Hill, NC.
2   We are happy to report that the Pittsburgh inner-city schools are well above average in maintenance. Still the district's budget is clearly stretched to a crisis point in maintaining academic programs at the same time it responds to federal and state mandates.
3   For further information, contact Each One-Reach One, Los Angeles Unified School District, 450 North Grand Avenue, Building G-253, Los Angeles, CA 90012.
4   The information on the Job Corps in this section is drawn from secondary sources, indicated in the text, but also printed materials distributed by the Pittsburgh Center as well as based on interviews of some staff members.

Appendix: Early School Leavers Protocol

Department of Administrative and Policy Studies
School of Education
University of Pittsburgh

NAME _____

ADDRESS _____

_____

PHONE _____

I   **Biographical Background**

    A   When were you born?
    B   Where were you born?
    C   Where do you live? Did your family always live there?
    D   What family members live in your household?
    E   Where do your parent(s) guardian(s) work? How long? Former employment?
    F   Do/did you have responsibilities at home?

II   **In-school Factors**

    A   *Teachers*

        1   Who were your good teachers? Why? Example?
        2   Who were your poorest teachers? Why? Example?

3 Did you always understand what teachers said to you?
4 Did your teachers care about students?
5 Did your teachers have favorites? If so, who? Give examples.

B *Administrators*

1 Do you remember the Building Principal at the last school you attended? What did he/she do? Example?
2 Do you remember the Vice-Principal at the last school you attended? What did he/she do? Example?
3 Does any other administrator stand out in your mind?
4 Did you have any trouble at school? What came of it? Were you treated fairly?
5 What did counselors do in your school? Do you recall talking with any counselors? If so, what about?
6 Did your school(s) have a social worker? What did the social worker do?
7 What do you remember about the people who worked in the school office? What did they do?
8 Did your school(s) have a security force? What did they do? Did you have any problems with them? Were they fair? Were they necessary?

C *Peers*

1 How did you get along with other students? If not well, why?
2 Who were your friends? Why were these your friends?
3 What did you do with your friends in school?
4 Did you and your friends ever get into trouble in school? If so, why? If not, why?

D   *Curriculum*

    1 Which of your subjects did you like most? Why?
    2 Which of your subjects did you like least? Why?
    3 Did you ever take part in the school's activity program? If yes, which ones? Why? What did you think of them?
    4 Did you ever go on any field trips while you were in school? What did you think of them?

E   *School Environment*

    1 How large was your school?
    2 Did you feel safe in the school? If no, why not?
    3 Was the school clean?
    4 Did you ride the bus or walk to school? Any problems?
    5 Did you ever feel discriminated against? If so, by whom? Describe the incident.
    6 Was the school building well kept?
    7 Was there a lot of noise?
    8 Was your school a good place to learn? If not, why?

III   Out-of-school Factors

A   *Media*

    1 Television
       (a) Do you watch television much?
       (b) What do you like to watch most on TV? Why?
       (c) About how many hours a day do you watch TV? Why?
    2 Printed matter
       (a) Do you read newspapers? Which?
       (b) Do you read magazines? Which?
       (c) Do you read books? What kind? Most recent?
    3 Films
       (a) What kinds of films do you like?
       (b) What have you seen most recently?

B   *Entertainment — what do you like to do for fun?*

C   *Work*

1   Are you working now?
2   Have you ever had other jobs? Describe it (them) to me.
3   When did you start working? Why did you work?
4   Did you enjoy your job(s)? Why?
5   Did it interfere with school? How?
6   Which did you enjoy more, school or work? Why?
7   Which do you think is more valuable, school or work? Why?

D   *Racial Discrimination*

Does it make a difference to be black or white in this country? If so, how does it make a difference?

E   *Neighborhood*

1   Did you spend a lot of time in your neighborhood? If yes, where? And with whom? And what did you do?
2   What did you like about your neighborhood? What didn't you like?
3   Was there a Boys' Club or Y-Group in your community? Church youth group? Organized sports? Did you belong to any of these?
4   What did you do with your friends outside of school?

F   *Mobility*

1   How often do you get outside of your neighborhood?
2   Do you get out of Pittsburgh often? If yes, where do you go? How? Why?

IV Follow-up Questions

A Why did you drop out of school?

B If you had it to do all over again, would you drop-out? Why? Why not?

C What could have been different about school that would have kept you there?

D Would you like to return to school? Do you *think* (or *plan*) you will return?

E What do you think about the kids who finished school?

F From what you have told me, what was the worst thing about school? Best?

INTERVIEW CONSENT FORM

Early School Leavers Project

You have been asked for information to be used in connections with the student dropout project. A tape recording of your interview will be made. This will also be converted to a typescript. In addition, your school records will be examined by the interviewer or principal investigators for the project, Dr. David E. Engel, Dr. Richard J. Altenbaugh and Dr. Don T. Martin. Your identity will be completely protected regarding any use of any of this material.

*   *   *

I understand that any information about me obtained for this research will be kept strictly confidential. Such information will not carry any personal identifying material and will be kept in locked files. It has also been explained to me that my identity will not be revealed in any description or publication of this research.

*   *   *

I also understand that I am free to refuse to participate in this study or to end my participation at any time.

*   *   *

I certify that I have read the preceding or it has been read to me and I understand its contents. A copy of this consent form will be given to me. My signature below means that I have freely agreed to participate in this study and to have my school records examined.

_____   _____
Date   Interviewee (Signature)

*   *   *

I certify that I have explained to the above individual the nature and purpose of this research study, have answered any questions that have been raised, and have witnessed the above signature.

_____   _____
Date   Interviewer (Signature)

For minors, 17 years of age and under:

I have also read the preceding and agree to the participation of my child.

_____   _____
Date   Parent/Guardian

                                                              _____
                                                              Witness

References

ALEXANDER, O.K.L., NATRIELLO, G. and PALLAS, A.M. (1985) 'For whom the school bell tolls: The impact of dropping out on cognitive performance', *American Sociological Review*, **50**, June, pp. 409–20.

ALTENBAUGH, R.J. (1981) 'Our children are being trained like dogs and ponies: Schooling, social control, and the working class', *History of Education Quarterly*, **21**, summer, pp. 213–22.

ALTENBAUGH, R.J. (1992) 'Teachers and the workplace' in ALTENBAUGH, R.J. (Ed) *The Teachers' Voice: A Social History of Teaching in Twentieth Century America*, London, Falmer Press.

ALTENBAUGH, R.J. (1993) 'Families, children, schools, and the workplace' in ROTHSTEIN, S. (Ed) *Handbook of Urban Education*, Westport, CT, Greenwood Press.

ANGUS, D.L. (1965) 'The dropout problem: An interpretive history', unpublished doctoral dissertation, Ohio State University.

ANGUS, D.L. and MIREL, J.E. (1985) 'From spellers to spindles: Workforce entry by the children of textile workers, 1888–1890', *Social Science History*, **9**, spring, pp. 123–44.

ANYON, J. (1979) 'Ideology and United States history textbooks', *Harvard Educational Review*, **49**, August, pp. 361–86.

ANYON, J. (1983) 'Workers, labor and economic history, and textbook content' in APPLE, M.W. and WEIS, L. (Eds) *Ideology and Practice in Schooling*, Philadelphia, PA, Temple University Press.

APPLE, M.W. (1986) *Teachers and Texts: A Political Economy of Class and Gender Relations in Education*, New York, Routledge and Kegan Paul.

APPLE, M.W. (1989) 'American realities: Poverty, economy, and education' in WEIS, L., FARRAR, E. and PETRIE, H.G. (Eds) *Dropouts from School: Issues, Dilemmas, and Solutions*, Albany, State University of New York Press.

ARIES, P. (1962) *Centuries of Childhood: A Social History of Family Life*, (trans Baldick, R.) New York, Vintage Books.

ATKINSON, P. (1985) *Language, Structure, and Reproduction*, London, Methuen.

BACHMAN, J.G., O'MALLEY, P.M. and JOHNSTON, J. (1979) 'Excerpt from *Adolescence to Adulthood*', *Educational Leadership*, **36**, April, pp. 481–6.

BALFOUR, M.H. and HARRIS, L.H. (1979) 'Middle-class dropouts: Myths and observations', *Education Unlimited*, **1**, April, pp. 12–16.

BAUM, W.K. (1981) *Transcribing and Editing Oral History*, Nashville, TN, American Association for State and Local History.

BECK, L. and MUIA, J.A. (1980) 'A portrait of a tragedy: Research findings on the dropout', *High School Journal*, **64**, November, pp. 65–72.

BECKER, H.S. (1951/52) 'Social-class variations in the teacher–pupil relationship', *Journal of Educational Sociology*, **25**, pp. 451–65.

BENNETT, K.P. and LECOMPTE, M.D. (1990) *The Way Schools Work: A Sociological Analysis of Education*, New York, Longman.

BICKEL, W.E., BOND, L. and LEMAHIEU, P. (1986) *Students At-Risk of not Completing High School*, Background Report to the Pittsburgh Foundation, RA.

BODNAR, J. (1982) *Workers' World: Kinship, Community, and Protest in An Industrial Society, 1900–1940*, Baltimore, MD, Johns Hopkins University Press.

BODNAR, J. (1985) *The Transplanted: A History of Immigrants in Urban America*, Bloomington, IN, Indiana University.

BOGDAN, R.C. and BIKLEN, S.K. (1982) *Qualitative Research for Education: An Introduction to Theory and Methods*, Boston, MA, Allyn and Bacon.

BORMAN, K.M. and SPRING, J.H. (1984) *Schools in Central Cities: Structure and Process*, New York, Longman Press.

BORUS, M.E. and CARPENTER, S.A. (1983) 'A note on the return of dropouts to high school', *Youth and Society*, **14**, June, pp. 501–7.

BOURDIEU, P. (1977) 'Cultural reproduction and social reproduction' in KARABEL, J. and HALSEY, A.H. (Eds) *Power and Ideology in Education*, New York, Oxford University Press.

BOURDIEU, P. and PASSERON, J.C. (1977) *Reproduction in Education, Society and Culture*, London, Sage Publications.

BOWLES, S. and GINTIS, H. (1976) *Schooling in Capitalist America: Educational Reform and the Contradictions of Economic Life*, New York, Basic Books.

BOYER, E.L. (1983) *High School: A Report on Secondary Education in America*, New York, Harper and Row.

BOYER, R.O. and MORAIS, H.M. (1976) *Labor's Untold Story*, New York, United Electrical, Radio and Machine Workers of America.

BRACEY, G.W. (1991) 'The condition of public education', *Phi Delta Kappan*, 74, October, pp. 105–17.

BRACEY, G.W. (1992) 'The second Bracey report on condition of public education', *Phi Delta Kappan*, 93, October, pp. 104–17.

BROPHY, J. (1985) 'Interactions of male and female students with male and female teachers' in WILKINSON, L.C. and MARRETT, C.B. (Eds) *Influences in Classroom Interaction*, Orlando, FL, Academic Press.

BRYK, A.S. and THUM, Y.M. (1989) 'The effects of high school organization on dropping out: An exploratory investigation', *American Educational Research Journal*, 26, Fall, pp. 353–83.

BURGESS, C. (1976) 'The goddess, the school book, and compulsion', *Harvard Educational Review*, 46, May, pp. 199–216.

BYINGTON, M. (1910; 1974) *Homestead: The Households of a Mill Town*, Pittsburgh, PA, University of Pittsburgh.

CALLAHAN, R.E. (1962) *Education and the Cult of Efficiency: A Study of the Social Forces that Have Shaped the Administration of the Public Schools*, Chicago, IL, University of Chicago Press.

CHASE-LANSDALE, P.L. and VINOVSKIS, M. (1993) 'Adolescent pregnancy and child support' in WOLLONS, R. (Ed) *Children At Risk in America: History, Concepts, and Public Policy*.

CLAYBAUGH, G.K. (1992) 'Textbooks and true believer', *Educational Horizons*, 70, summer, pp. 159–61.

CLEMENTS, B.S. (1990) 'What is a dropout? Pilot program collects meaningful data for improving schools', *School Administrator*, 47, March, pp. 18–22.

CLIFFORD, G.J. (1978) 'Home and school in 19th-century America: Some personal-history reports from the United States', *History of Education Quarterly*, 18, spring, pp. 3–34.

COHEN, S. (1968) 'The industrial education movement, 1906–1917', *American Quarterly*, 20, pp. 95–110.

COHEN, S. (1973) *Education in the United States*, New York, Random House.

COLLISTER, L. (1994) 'True commitment through each one-reach one', *The National Dropout Prevention Newsletter*, 7, summer.

COMMISSION ON READING (1984) *Becoming a Nation of Readers*, Washington, DC, National Institute of Education.

CONRATH, J. (1984) 'Snatching victory from the jaws of learning defeat: How one school fought the dropout blitz', *Contemporary Education*, 56, Fall, pp. 36–8.

CONZEN, K.N. (1976) *Immigrant Milwaukee, 1836–1860: Accommodation and Community in a Frontier City*, Cambridge, MA, Harvard University Press.

COUNTS, G.S. (1922) *The Selective Character of American Secondary Education*, New York, Arno Press.

CREMIN, L.A. (1965) *The Wonderful World of Ellwood Patterson Cubberley: An Essay on the Historiography of American Education*, New York, Teachers College Press.

CREMIN, L.A. (1980) *American Education: The National Experience, 1783–1876*, New York, Harper and Row.

CREMIN, L.A. (1988) *American Education: The Metropolitan Experience, 1876–1980*, New York, Harper and Row.

CUBAN, L. (1984) *How Teachers Taught: Constancy and Change in American Classrooms, 1890–1980*, New York, Longman Inc.

CUTLER, W. (1971) 'Oral history — its nature and uses for educational history', *History of Education Quarterly*, **11**, summer, pp. 184–94.

DAVIS, C. (1977) *Oral History: From Tape to Tape*, Chicago, American Library Association.

DEWEY, J. (1963) *Experience and Education*, New York, Collier Books.

DOHRENWEND, B.S. and RICHARDSON, S.A. (1964) 'A use for leading questions in research interviewing', *Human Organization*, **23**, pp. 76–7.

DORN, S. (1993) 'Origins of the "dropout problem"', *History of Education Quarterly*, **33**, fall, pp. 353–74.

DUBLIN, T. (1979) *Women at Work: The Transformation of Work and Community in Lowell, Massachusetts, 1826–1860*, New York, Columbia University Press.

EKSTROM, R.B., GOERTZ, M.E., POLLACK, J.M. and ROCK, D.A. (1986) 'Who drops out of high school and why? Findings from a national study', *Teachers College Record*, **87**, spring, pp. 356–73.

ELLIOTT, D.S., VOSS, H.L. and WENDLING, A. (1966) 'Capable dropouts and the social milieu of the high school', *Journal of Educational Research*, **60**, December, pp. 180–6.

ELLIS, H.C. (1903) 'The percentage of boys who leave the high school and the reasons therefor', *National Education Association Journal of Proceedings and Addresses*, pp. 792–801.

ENGEL, D.E. (1994) 'School leavers in American society: Interviews with school dropouts/dropbacks' in MORRIS, R.C. (Ed) *Using What We Know About At-Risk Youth: Lessons from the Field*, Lancaster, PA: Technomic Publishing Co, Inc.

ENSIGN, F.C. (1969) *Compulsory School Attendance and Child Labor: A Study of the Historical Development of Regulations Compelling Attendance and Limiting the Labor of Children in a Selected Group of States*, New York, Arno Press (original work published in 1921).

EUSTIS, T.W. (1976) 'Get it in writing: Oral history and the law', *Oral History Review*, pp. 6–18.

FARRELL, E. (1990) *Hanging In and Dropping Out: Voices of At-Risk High. Students*, New York, Columbia University Press.

FEINBERG, W. and SOLTIS, J.F. (1985) *School and Society*, New York, Teachers College Press.

FENSHAM, P. (Ed) (1986) *Alienation from Schooling*, London, Routledge and Kegan Paul.

FINE, M. (1985) 'Dropping out of high school: An inside look', *Social Policy*, **16**, fall, pp. 43–50.

FINE, M. (1986) 'Why urban adolescents drop into and out of public high school', *Teachers College Record*, **87**, spring, pp. 393–409.

FINE, M. (1991) *Framing Dropouts: Notes on the Politics of an Urban Public High School*, Albany, NY, State University of New York Press.

FINE, M. and ZANE, N. (1989) 'Bein' wrapped too tight: When low-income women drop out of high school' in WEIS, L., FARRAR, E., and PETRIE, H.G. (Eds) *Dropouts from School: Issues, Dilemmas, and Solutions*, Albany, NY, State University of New York Press.

FINKELSTEIN, B.J. (1976) 'In fear of childhood: Relationships between parents and teachers in popular primary schools in the nineteenth century', *History of Childhood Quarterly*, **3**, winter, pp. 321–35.

FINKELSTEIN, B.J. (1991) 'Dollars and dreams: Classrooms as fictitious message systems, 1790–1930', *History of Education Quarterly*, **31**, winter, pp. 463–88.

FIRESTONE, W.A. and ROSENBLUM, S. (1988) 'Building commitment in the urban high schools', *Educational Evaluation and Policy Analysis*, **10**, winter, pp. 285–99.

FISHER, I. (1992) 'The comeback trail', *New York Times*, Section 4A, 2 August, p. 21.

FOLEY, E. (1985) 'What will recalibrated standards hold for dropout-prone students?', *Education Week*, **17** April, pp. 17 and 24.

FORDHAM, S. and OGBU, J.U. (1986) 'Black students' school success: Coping with the "burden of 'acting White'"', *Urban Review*, **18**, 3, pp. 176–206.

FROMM, E. (1955) *The Same Society*, New York, Fawcett World Library.

GADWA, K. and GRIGGS, S.A. (1985) 'The school dropout: Implications for counselors', *School Counselor*, **33**, September, pp. 9–17.

GENERAL ACCOUNTING OFFICE (1987) *School Dropouts: Survey of Local Programs* (GAO/HRD-87-108) Washington, DC, Government Printing Office, July.

GIROUX, H. (1983) 'Theories of reproduction and resistance in the new sociology of education: A critical analysis', *Harvard Educational Review*, **53**, August, pp. 257.

GLAZIER, W. (1883) 'The great furnace of America' in LUBOVE, R. (Ed) (1976) *Pittsburgh: A Documentary History*, New York, New Viewpoints.

GOLDMAN, J.P. (1990) 'Some local districts struggle to adopt dropout standards', *School Administrator*, March, pp. 20–1.

GOODLAD, J.I. (1984) *A Place Called School: Prospects for the Future*, New York, McGraw-Hill Company.

GOULD, E.R.L. (1893) *The Social Condition of Labor*, Baltimore, MD, Johns Hopkins University Press.

GREENWOOD, J.M. (1900) 'Report on high school statistics', *National Education Association Journal of Proceedings and Addresses*, pp. 340–51.

GROSS, J. (1992) 'Remnant of the war on poverty, Job Corps is still a quiet success', *New York Times*, Section A, February 17, pp. 1 and 14.

GRUMET, J.R. (1988) *Bitter Milk: Women and Teaching*, Amherst, MA, University of Massachusetts Press.

HAMMACK, F.J. (1986) 'Large school systems' dropout reports: An analysis of definitions, procedures, and findings', *Teachers College Record*, **87**, spring, pp. 324–41.

HAMMARBERG, M.A. (1971) 'Designing a sample from incomplete historical lists', *American Quarterly*, **23**, pp. 542–61.

HAMOVITCH, B. (1993) Caring as an antidote to non-caring: Theoretical musings and empirical findings. Paper presented at AESA, Chicago, November.

HAREVEN, T.K. (1982) *Family Time and Industrial Time: The Relationship Between the Family and Work in a New England Industrial Community*, New York, Cambridge University Press.

HARRIS, R. (1975) *The Practice of Oral History: A Handbook*, Glen Rock, Microfilming Corporation of America.

HARRIS, W.T. (1873) 'The early withdrawal of pupils from school: Its causes and its remedies', *National Education Association Journal of Proceedings and Addresses*, pp. 260–73.

HAWES, J.M. (1971) *Children in Urban Society: Juvenile Delinquency in Nineteenth-Century America*, New York, Oxford University Press.

HAWES, J.M. (1991) *The Children's Rights Movement: A History of Advocacy and Protection*, Boston, Twayne Publishers.

HINER, N.R. (1983) 'Domestic cycles: History of childhood and family' in BEST, J.H. (Ed) *Historical Inquiry in Education: A Research Agenda*, Washington, DC, American Educational Research Association.

HINER, N.R. and HAWES, J.M. (1985) *Growing Up in America: Children in Historical Perspective*, Urbana, IL, University of Illinois Press.

HOGAN, D. (1978) 'Education and the making of the Chicago working class, 1880–1930', *History of Education Quarterly*, **18**, fall, pp. 227–70.

HOLLINGSHEAD, A.B. (1975) *Elmtown's Youth and Elmtown Revisited*, New York, John Wiley & Sons.

HORAN, P.M. and HARGIS, P.G. (1991) 'Children's work and schooling in the late nineteenth-century family economy', *American Sociological Review*, **56**, October, pp. 583–96.

HUNT, T.C. and CLAWSON, E.U. (1975) 'Dropouts: Then and now', *High School Journal*, **58**, March, pp. 237–50.

HURT, J.S. (1979) *Elementary Schooling and the Working Classes, 1860–1918*, London, Routledge and Kegan Paul.

ISSEL, W.H. (1967) 'Teachers and educational reform during the Progressive Era: A case study of the Pittsburgh Teachers Association', *History of Education Quarterly*, **7**, December, pp. 220–33.

JONES, E. F., et al (1986) *Teenage Pregnancy in Industrialized Countries: A Study Sponsored by the Alan Guttmacher Institute*, New Haven, CT, Yale University Press.

JONES, S. (1826) *Pittsburgh in the Year 1826*, New York, Arno Press.

KAESTLE, C.F. (1983) *Pillars of the Republic: Common Schools and American Society, 1780–1860*, New York, Hill and Wang.

KARABEL, J. and Halsey, A.H. (Eds) (1977) *Power and Ideology in Education*, New York, Oxford University Press.

KATZ, M.B. and HOGAN, D. (1983) 'Schools, work and family life: Social history' in BEST, J.H. (Ed) *Historical Inquiry in Education: A Research Agenda*, Washington, DC, American Educational Research Association, pp. 282–304.

KELLY, D.M. (1993) *Last Chance High: How Girls and Boys Drop in and Out of Alternative Schools*, New Haven, CT, Yale University Press.

KENNARD, B. (1914) 'Children's play' in LUBOVE, R. (Ed) (1976) *Pittsburgh: A Documentary History*, New York, New Viewpoints.

KENNISTON, K. (1965) *The Uncommitted: Alienated Youth in American Society*, New York, Harcourt, Brace and World.

KERCHNER, C.T. and KOPPICH, J.E. (1993) *A Union of Professionals: Labor Relations and Educational Reform*, New York, Teachers College Press.

KETT, J. (1977) *Rites of Passage: Adolescence in America, 1790 to the Present*, New York, Basic Books.

KOLSTAD, A.J. and OWINGS, J.A. (1986) *High School Dropouts Who Change Their Minds about School*, Washington, DC, Office of Educational Research and Improvement, Department of Education.

KOMINSKI, R. (1990) 'Estimating the national high school dropout rate', *Demography*, **27**, 2, May, pp. 303–11.

KOWALSKI, C.J. and CANGEMI, J.P. (1974) 'High school dropouts — A lost resource', *College Student Journal*, 8, November–December, pp. 71–4.

KOZOL, J. (1991) *Savage Inequalities: Children in America's Schools*, New York, Crown Publishers.

KUNISAWA, B.N. (1988) 'A nation in crisis: The dropout dilemma', *NEA Today*, January, pp. 61–5.

LEE, V.E. and EKSTROM, R.B. (1987) 'Student access to guidance counseling in high school', *American Educational Research Journal*, **24**, summer, pp. 287–310.

LEVIN, H.M. (1982) *Education and Work*, Program Report No. 82-B8, Stanford, CA, Institute for Research on Educational Finance and Governance.

LIVESAY, H.C. (1975) *Andrew Carnegie and the Rise of Big Business*, Boston, MA, Little, Brown and Company.

LOKEN, J.O. (1973) *Student Alienation and Dissent*, Englewood Cliffs, NJ, Prentice-Hall.

LORANT, S. (1964) *Pittsburgh: The Story of an American City*, Garden City, NY, Doubleday and Company, Inc.

LUBOVE, R. (Ed) (1976) *Pittsburgh: A Documentary History*, New York, New Viewpoints.

LYKE, R.F. (1987) *High School Dropouts* Washington, DC, Congressional Research Service. (Order Code IB7167)

McBRIDE, P.W. (1974) 'The Co-op industrial education experience, 1900–1917', *History of Education Quarterly*, **14**, summer, pp. 209–22.

McCOY, W.D. (1951) 'Public education in Pittsburgh, 1835–1950', *Western Pennsylvania Historical Magazine*, December, pp. 219–38.

MANN, D. (1986) 'Can we help dropouts: Thinking about the doable', *Teachers College Record*, **87**, spring, pp. 307–23.

MARKEY, J.P. (1988) 'The labor market problems of today's high school dropouts', *Monthly Labor Review*, **111**, June, pp. 36–43.

MERTON, R.K. (1938) 'Social structure and anomie', *American Sociological Review*, **3**, October, pp. 672–82.

MIREL, J.E. (1991) 'Adolescence in twentieth-century America' in LERNER, R.M., PETERSON, A.C. and BROOKS-GUNN, J. (Eds) *Encyclopedia of Adolescence*, Vol. II, New York, Garland Publishing, Inc.

MIREL, J.E. and ANGUS, D. (1985) 'Youth, work, and schooling in the Great Depression', *Journal of Early Adolescence*, **5**, pp. 489–504.

MORROW, G. (1986) 'Standardizing practice in the analysis of school dropouts', *Teachers College Record*, **87**, spring, pp. 342–55.

NASAW, D. (1985) *Children of the City: At Work and At Play*, New York, Oxford University Press.

NATIONAL CENTER FOR EDUCATIONAL STATISTICS (1992) *Dropout Rates in the United States: 1991*, Washington, DC, US Department of Education, Office of Educational Research and Improvement.

NATIONAL CENTER FOR EDUCATIONAL STATISTICS (1994) *Dropout Rates in the United States: 1993*, Washington, DC, US Department of Education, Office of Educational Research and Improvement.

NATIONAL COMMISSION OF EXCELLENCE IN EDUCATION (1983) *A Nation at Risk*, Washington, DC, US Government Printing Office.

THE NATIONAL TESTING DEBATE (1991) *Education Network News: National Council of La Raza*, **10**, July, pp. 1–4.

NATRIELLO, G. (1986) 'School dropouts: Patterns and policies', *Teachers College Record*, **87**, spring, pp. 305–6.

NATRIELLO, G., PALLAS, A.M. and McDILL, E.L. (1986) 'A population at risk: Potential consequences of tougher school standards for student dropouts', *American Journal of Education*, **94**, pp. 135–181.

NEIGHBORHOODS CLASH IN SCHOOL (1993) *Pittsburgh Post-Gazette*, 8 July, Section B, p. 1.

NEWMANN, F.M. (1991) 'Linking restructuring to authentic student achievement, *Phi Delta Kappan*, **72**, February, pp. 458–63.

NEWMANN, F.M. and WEHLAGE, G.G. (1993) 'Five standards of authentic instruction', *Educational Leadership*, **50**, April, pp. 8–12.

NEWMANN, F.M. WEHLAGE, G.G. and LAMBORN, S.D. (1992) 'The significance and sources of student engagement' in NEWMANN, F.M. (Ed) *Student Engagement and Achievement in American Secondary Schools*, New York, Teachers College Press.

NODDINGS, N. (1984) *Caring: A Feminine Approach to Ethics and Moral Education*, Berkeley, CA, University of California Press.

OBLINGER, C. (1978) *Interviewing the People of Pennsylvania: A Conceptual Guide to Oral History*, Harrisburg, PA, Pennsylvania Historical and Museum Commission.

OFFICE OF EDUCATIONAL RESEARCH AND IMPROVEMENT (OERI) (1987) *Dealing with Dropouts: The Urban Superintendents Call to Action*, Washington, DC, Government Printing Office.

O'NEIL, J. (1993) 'Achievement of US students debated', *Update: Association for Supervision and Curriculum Development*, March, pp. 3–4.

PHILLIPS, D.C. and SOLTIS, J.F. (1985) *Perspectives on Learning*, New York, Teachers College Press.

PITTMAN, R.B., and HAUGHWOUT, P. (1987) 'Influence of high school size on dropout rate', *Educational Evaluation and Policy Analysis*, **9**, winter, pp. 337–43.

POWELL, A.G., FARRAR, E. and COHEN, D.K. (1985) *The Shopping Mall High School: Winners and Losers in the Educational Marketplace*, Boston, MA, Houghton Mifflin Company.

RAWLS, J. (1971) *A Theory of Justice*, Cambridge, MA, Belknap Press.

REID, K. (1983) 'Retrospection and persistent school absenteeism', *Educational Research*, **25**, pp. 110–15.

REVEL, J.F. (1971) *Without Marx or Jesus: The American Revolution Has Begun* (trans. Bernard, J.) New York, Doubleday.

'Riley announces "Goals 2000: Educate America Act"' (1993) *Community Update: US Department of Education*, May, pp. 1–4.

RIPPA, S.A. (1976) *Education in a Free Society: An American History*, New York, David McKay.

RIST, R.C. (1970) 'Student social class and teacher expectations: The self-fulfilling prophecy in ghetto education', *Harvard Educational Review*, **40**, pp. 411–51.

RODERICK, M. (1993) *The Path to Dropping Out: Evidence for Intervention*, Westport, CT, Auburn Press.

ROGERS, C.R., and FREIDBERG, H.J. (1994) *Freedom to Learn*, New York, Merrill Publishing Co.

ROMNEY, J. (1973) 'Legal considerations in oral history', *Oral History Review*, pp. 191–208.

RORABAUGH, W.J. (1986) *The Craft Apprentice: From Franklin to the Machine Age in America*, New York, Oxford University Press.

ROZYCKI, E.G. (1992) 'The textbook: Tool or symbol?' *Educational Horizons*, **70**, summer, pp. 161–3.

RUMBERGER, R.W. (1983) 'Dropping out of high school: The influence of race, sex, and family background', *American Educational Research Journal*, **20**, summer, pp. 199–220.

RUMBERGER, R.W. (1986) 'High school dropouts: A review of issues and evidence', *Review of Education Research*, **57**, summer, pp. 101–21.

RUMBERGER, R.W. (1991) 'Chicano dropouts: A review of research and policy issues' in VALENCIA, R.R. (Ed) *Chicano School Failure and Success: Research and Policy Agendas for the 1990s*, London, Falmer Press.

SCHENLEY SHOOTING SUSPECT ARRESTED (1993) *Pittsburgh Post-Gazette*, Section B, 8 July 1993, p. 1.

SEEMAN, M. (1959) 'On the meaning of alienation', *American Sociological Review*, **24**, December, pp. 847–89.

SEEMAN, M. (1971) 'Alienation: A map', *American Sociological Review*, **5**, August, pp. 82–95.

SEIDMAN, I.E. (1991) *Interviewing as Qualitative Research: A Guide for*

Researchers in Education and the Social Sciences, New York, Teachers College Press.

SIZER, R. (1984) *Horace's Compromise: The Dilemma of the American High School*, Boston, MA, Houghton Mifflin.

SOLOMON, R.P. (1989) 'Dropping out of academics: Black youth and the sports subculture in a cross-national perspective' in WEIS, L., FARRAR, E. and PETRIE, H.G. (Eds) *Dropouts from School: Issues, Dilemmas, and Solutions*, Albany, NY, State University of New York Press.

SPRADLEY, J.P. (1979) *The Ethnographic Interview*, New York, Holt, Rinehart and Winston.

SPRING, J. (1976) *The Sorting Machine: National Educational Policy Since 1945*, New York, David MacKay.

SPRING, J. (1991) *American Education: An Introduction to Social and Political Aspects*, New York, Longman Press.

STARR, L.M. (1977) 'Oral history', *Encyclopedia of Library and Information Science*, **20**, New York, Marcel Dekker.

STEVENSON, H.W. (1993) 'Bracey's broadsides are unfounded', *Educational Leadership*, **50**, February, p. 60.

STONE, A. (1993) 'Kids, guns: "It's shoot or be shot"', *USA Today*, June 3, pp. 1 and 2.

STOUGHTON, C.R. and GRADY, B.R. (1978) 'How many students will drop out and why?', *North Central Association Quarterly*, **53**, pp. 312–5.

STRICKLAND, C.E. and BURGESS, C. (Eds) (1965) *Health, Growth, and Heredity: G. Stanley Hall on Natural Education*, New York, Teachers College Press.

SWARTZ, D. (1977) 'Pierre Bourdieu: The cultural transmission of social inequality', *Harvard Educational Review*, **47**, November, pp. 545–55.

'TEXT OF STATEMENT OF EDUCATION ADOPTED BY GOVERNORS' (1990) *Education Week*, 7 March, pp. 16–17.

'Tracking progress toward the national goals' (1991) *OERI Bulletin* (Office of Educational Research and Improvement, US Department of Education), p. 4.

THOMPSON, E.P. (1967) 'Time, work-discipline, and industrial capitalism', *Past and Present*, **38**, pp. 56–97.

THOMPSON, P. (1978) *The Voice of the Past: Oral History*, New York, Oxford University Press.

TODD, H.M. (1912/13) 'Why children work: The children's answer', *McClure's Magazine*, **40**, Nov.–Apr., pp. 68–79.

TYACK, D. (1976) 'Ways of seeing: An essay on the history of compulsory schooling', *Harvard Educational Review*, **46**, August, pp. 355–89.

TYACK, D., and BERKOWITZ, M. (1977) 'The man nobody liked: Toward a social history of the truant officer, 1840–1940', *American Quarterly*, **29**, spring, pp. 31–54.

TYACK, D., and HANSOT, E. (1990) *Learning Together: A History of Coeducation in American Public Schools*, New Haven, CT, Yale University Press.

US DEPARTMENT OF LABOR (1911) *Report on the Condition of Women and Child Wage Earners in the United States: Employment of Women in Metal Trades*, Bureau of Labor Statistics, Volume II, Washington, DC, Government Printing Office.

US DEPARTMENT OF LABOR (1916) *Summary of the Report on Conditions of Women and Child Wage Earners in the United States*. Bureau of Labor Statistics, Bulletin No. 175, Washington, DC, US Government Printing Office.

US DEPARTMENT OF LABOR (1990) 'Job Corps: Analysis of costs invested in human capital in the Job Corps program by regional office', Office of Inspector General, Report No. 12–91–034–03–370.

US SENATE (1885) *Report Upon the Relations Between Labor and Capital: Volume I and II*, Washington, DC, Government Printing Office.

VINOVSKIS, M. (1986) 'Teenage pregnancy', *Social Science*, **71**, fall, pp. 158–64.

VIOLAS, P.C. (1978) *The Training of the Urban Working Class: A History of Twentieth-Century American Education*, Chicago, IL, McNally.

VOSS, H.L., WENDLING, A. and ELLIOTT, D.S. (1966) 'Some types of high school dropouts', *Journal of Educational Research*, **59**, April, pp. 36–367.

WAGNER, H. (1984) 'Why poor kids quit attending school', *Education*, **105**, pp. 185–8.

WALTERS, P.B. and O'Connell, P.J. (1988) 'The family economy, work and education participation in the United States, 1890–1940', *American Journal of Sociology*, **93**, March, pp. 1116–52.

WEHLAGE, G.G. (1989) 'Dropping out: Can schools be expected to prevent it?' in WEIS, L., FARRAR, E. and PETRIE, H.G. (Eds) *Dropouts from School: Issues, Dilemmas, and Solutions*, Albany, NY, State University of New York Press.

WEHLAGE, G.G. and RUTTER, R.A. (1986) 'Dropping out: How much do schools contribute to the problem?' *Teachers College Record*, **87**, spring, pp. 374–93.

WEHLAGE, G.G., RUTTER, R.A., SMITH, G.A., LESKO, N. and FERNANDEZ,

R.R. (1989) *Reducing the Risk: School as Communities of Support*, London, Falmer Press.

WEIDMAN, J.C. and FRIEDMANN, R.R. (1984) 'The school-to-work transition for high school dropouts', *Urban Review*, **16**, pp. 25–42.

WEIS, L., FARRAR, E. and PETRIE, H.G. (Eds) *Dropouts from School: Issues, Dilemmas and Solutions*, Albany, NY, SUNY Press.

WELTER, B. (1966) 'The cult of true womanhood: 1820–1860', *American Quarterly*, 18, summer, pp. 151–74.

WEST, C. (1993a) *Race Matters*, Boston, MA, Beacon Press.

WEST, C. (1993b) 'We must endure lovingly', *Harvard Divinity Bulletin*, 1, summer, pp. 1–2.

WILLIS, P. (1977) *Learning to Labor: How Working-Class Kids Get Working-Class Jobs*, New York, Columbia University Press.

WIRTH, A.G. (1983) *Productive Work in Industry and Schools: Becoming Persons Again*, Lanham, MD, University Press of America.

YANS-MCLAUGHLIN, V. (1971) 'Patterns of work and family organization: Buffalo's Italians', *Journal of Interdisciplinary History*, **2**, autumn, pp. 299–314.

YANS-MCLAUGHLIN, V. (1977) *Family and Community: Italian Immigrants in Buffalo, 1880–1930*, Ithaca, NY, Cornell University Press.

ZELIZER, V.A. (1985) *Pricing the Priceless Child: The Changing Social Value of Children*, New York, Basic Books.

Index

DATE DUE

MAY 1 0 1999	
OCT 2 0 1999	

BRODART
Cat. No. 23-221